Fundamentals of Microgrids

Fundamentals of Microgrids
Development and Implementation

Edited by
Stephen A. Roosa

CRC Press is an imprint of the
Taylor & Francis Group, an **informa** business

First edition published 2021
by CRC Press
6000 Broken Sound Parkway NW, Suite 300, Boca Raton, FL 33487-2742

and by CRC Press
2 Park Square, Milton Park, Abingdon, Oxon, OX14 4RN

© 2021 Taylor & Francis Group, LLC

CRC Press is an imprint of Taylor & Francis Group, LLC

Reasonable efforts have been made to publish reliable data and information, but the author and publisher cannot assume responsibility for the validity of all materials or the consequences of their use. The authors and publishers have attempted to trace the copyright holders of all material reproduced in this publication and apologize to copyright holders if permission to publish in this form has not been obtained. If any copyright material has not been acknowledged please write and let us know so we may rectify in any future reprint.

Except as permitted under U.S. Copyright Law, no part of this book may be reprinted, reproduced, transmitted, or utilized in any form by any electronic, mechanical, or other means, now known or hereafter invented, including photocopying, microfilming, and recording, or in any information storage or retrieval system, without written permission from the publishers.

For permission to photocopy or use material electronically from this work, access www.copyright.com or contact the Copyright Clearance Center, Inc. (CCC), 222 Rosewood Drive, Danvers, MA 01923, 978-750-8400. For works that are not available on CCC please contact mpkbookspermissions@tandf.co.uk

Trademark notice: Product or corporate names may be trademarks or registered trademarks, and are used only for identification and explanation without intent to infringe.

ISBN: 978-0-367-53539-1 (hbk)
ISBN: 978-1-003-08240-8 (ebk)

Typeset in Times
by Deanta Global Publishing Services, Chennai, India

Dedication

This book is dedicated to my three wonderful children, Sarah, Daniel, and Maryelle. They have grown into caring and responsible adults. I know that during their lifetimes, they and others like them will continue work to help ensure the sustainability of our planet. After all, we must focus our efforts on creating a better world for future generations.

Contents

Preface xv
Acknowledgments xvii
Editor Bio xix
Contributors xxi
List of Acronyms xxiii
Introduction xxvii

Chapter 1 Introduction to Microgrids 1

    What Are Microgrids? 1
        Standalone Power 2
        Distributed Energy Resources 3
        Why Would You Need a Microgrid? 4
        Types of Microgrids 6
        How Does a Microgrid Help? 8
    Advantages and Disadvantages of Microgrids 9
        Advantages of Microgrids 11
        Disadvantages of Microgrids 12
    Summary 13
    References 15

Chapter 2 Environmental Drivers for Microgrid Development 17

    Introduction 17
    The Hydrocarbon Age Has Arrived 18
        Carbon Dioxide Emissions 20
        International Policies—Kyoto and the Paris Agreement 22
        Carbon Management—Adapt or Mitigate? 24
        Costs of Reducing Greenhouse Gas Emissions 25
        Potential for Reducing Carbon Emissions 26
        The New Economics of Coal Plants 27
    Transitioning to a Low-Carbon Economy 28
        Greenhouse Gas Reduction Strategies 29
        Other Environmental Concerns 30
        Microgrids Are One Solution 32
    Conclusions 33
    References 34

Chapter 3 The Roots of Microgrids 37

    The Early History of Microgrids 37
        Hydropower Microgrids 37
        The World's Largest Microgrid in 1883 38

		Alternating Current 39
		Combined Heat and Power 39
		Mill Towns as a Model for Today's Microgrids 40
		Ships Are Microgrids 41
		Building-Scale Nanogrids 42
		Sustainable Buildings 42
		Green Building Assessment Methods 44
		Zero Net Energy Buildings 44
		Nanogrids 45
		Regional Grids 46
		The Role of Transportation Systems 48
		Summary 50
		References 51
Chapter 4	Traditional Electrical Supply Systems 53
	Supplying the Electric Grid 53
	What Is the Electric Grid? 53
	Categories of Energy Resources 54
	Traditional Sources of Grid-Supplied Elecrical Power 56
	Large Hydropower 56
	Coal-Fired Electrical Power 57
	Natural Gas-Fired Electrical Power 58
	Diesel Generation 59
	Nuclear Power 60
	Problems with the Electric Grid 62
	Security Issues 62
	System Inefficiencies and Power Outages 64
	Environmental Issues 65
	Regulation of the Electric Grid 65
	Connecting Microgrids with the Grid 66
	Summary 67
	References 68
Chapter 5	Microgrid Architecture and Regulation 71
	Features of Microgrid Architecture 71
	Microgrid Operational Configurations 71
	Planning for Electrical Distribution Systems 73
	Alternative Architectures for AC and DC Microgrids 75
	Components of Microgrids 76
	Power Sources 76
	Point of Common Coupling 77
	Microgrid Power Management Systems 77
	Categories of Loads 79
	Energy Storage Systems 79

Contents ix

	Advanced Microgrids	81
	Microgrid Regulations and Standards	84
	U.S. Clean Air Act	85
	U.S. Public Utility Regulatory Policies Act (PURPA)	85
	IEEE Standard 1547-2018 for Interconnection and Interoperability of Distributed Energy Resources with Associated Electric Power Systems Interfaces	86
	IEEE P1547.4-2011 Guide for Design, Operation, and Integration of Distributed Resource Island Systems with Electric Power Systems	87
	IEEE P2030.7-2007 IEEE Standard for the Specification of Microgrid Controllers	87
	IEC 61727 International Electrotechnical Commission's PV System Requirements	87
	Microgrid Standards Being Developed	88
	Summary	88
	Acknowledgments	89
	References	89
Chapter 6	Linking Microgrids with Renewable Generation	91
	The Impact of Renewable Energy	91
	Renewable Generation for Microgrids	93
	Small Hydropower Systems	94
	Biomass Energy	95
	Landfill Gas Extraction	95
	Solar Energy (Thermal)	97
	Solar Energy (Photovoltaic)	98
	Wind Power	100
	Geothermal Energy	100
	Waste-to-Energy	102
	Comparitive Cost of Generation Systems	103
	Summary	105
	References	106
Chapter 7	Energy Storage Technologies for Microgrids	109
	Energy Storage Solves Problems	109
	Electrical Energy Storage	110
	Mechanical Energy Storage	111
	Pumped Storage Hydroelectricity	111
	Compressed Air Energy Storage	114
	Flywheel Energy Storage	116
	Electrical/Electrochemical Energy Storage	116
	Lead-Acid Batteries	117
	Lithium-Ion Batteries	117

	Sodium Sulfur Batteries	118
	Vanadium Redox Batteries	119
	Zinc-Air Batteries	119
	Hydrogen Energy Storage	120
	Thermal Energy Storage	120
	Summary	121
	References	123
Chapter 8	Hybrid Generation Systems for Microgrids	125
	Energy Resources for Microgrids	125
	Hybrid Generation Systems	126
	What Are Hybrid Power Systems?	126
	Advantages of Hybrid Generation	128
	Examples of Hybrid Generation	129
	Diesel and Renewable Hybrid Systems	129
	Natural Gas and Renewable Energy Systems	129
	Solar and Wind Power	130
	Solar and Geothermal Energy	131
	Nuclear and Renewable Energy	133
	Fuel Cells and Renewable Hybrid Generation	134
	Renewable Co-Generation	135
	Summary	136
	References	137
Chapter 9	Community and Local Microgrids	141
	Community Microgrids	141
	Drivers for Community-Scale Microgrids	141
	U.S. Municipal Renewable Energy Goals	142
	Types of Microgrids	143
	Mobile Microgrids	144
	Local Service Microgrids	144
	Military Microgrids	145
	Industrial Microgrids	146
	Utility Distribution Microgrids	147
	Campus Microgrids	148
	Virtual Power Plants	148
	Examples of Community and Local Microgrids	150
	Kodiak Island	150
	Borrego Springs Microgrid	151
	Long Island Community	152
	Summary	153
	References	154

Contents   xi

Chapter 10 Ghana's Transition to Renewable Energy Microgrids 157

Ishmael Ackah, Eric Banye, Dramani Bukari, Eric Kyem, and Shafic Suleman

Introduction 157
    Access to Electricity in Africa 157
    Ghana's Renewable Energy Act 158
Microgrid Solar PV System Viability in Ghana 158
Perspectives on Microgrids 159
    Energy Requirements of Rural Communities 163
Methodology 164
Discussion of Findings 165
    Microgrid Policy and Regulation in Ghana 165
    Private and Government Microgrid Systems 165
    Capacity and Reliability 166
    Technologies 166
    Tariffs and Rates 167
Impact of Microgrid Development 167
    Impact on Expenditures for Fuel 167
    Impact on Women 168
    Impact on Education 168
Conclusions and Recommendations 168
    Conclusions 169
    Recommendations 170
Acknowledgments 170
References 171

Chapter 11 Local Energy Supply Possibilities—Islanding Microgrid Case Study 173

István Vokony, József Kiss, Csaba Farkas, László Prikler, and Attila Talamon

Introduction 173
International Overview 174
    Guaranteed Service Levels 176
    Construction Aspects 176
    Planning Principles 177
    Operational Aspects 178
Local Solution: Container Microgrid 179
    Container Microgrid Assessment 180
Decision Support for Software Development 182
Cost of Network Development Alternatives 183
Summary 185
Acknowledgments 185
References 185

Chapter 12 Energy Blockchain—Advancing DERs in Developing Countries 189

Alain G. Aoun

Introduction 189
Blockchain Opportunities for Energy Trading................................ 190
DER Challenges in Developing Countries 192
Incentives for Developing Countries 193
    Blockchain and Smart Meters 193
    Smart Contracts and P2P Energy Trading 194
    Energy Backed Currencies 195
    Blockchain for Electric Vehicle Charging..................................... 196
Key Challenges and Barriers... 197
Conclusion 198
Acknowledgments 199
References 199

Chapter 13 Smart Microgrids 201

Introduction 201
Smart Microgrids and Their Benefits... 201
    Advanced Microgrids 202
    Advanced Remote Microgrids... 202
        Mandatory Microgrid Features 202
        Preferred Microgrid Features 203
    Differences between Smart Grids and Smart Microgrids 203
    Benefits of Smart Microgrids 204
Technologies Used for Smart Microgrids 205
    Sensor Systems 205
    Advanced Metering Infrastructure... 206
    Technologies 206
        Residential 207
        Commercial 207
        Electricity Distribution 207
Examples of Smart Microgrids 207
    Projects Under Development 208
    Projects Deployed 209
Use of Renewable Energy Technologies 210
    Sustainable Smart Microgrids.. 210
    Managing Multiple Renewable Microgrids................................ 211
Policy Concerns.. 212
Conclusions 213
References 214

Chapter 14 Scoping the Business Case for Microgrids 217

Introduction 217
Business Case for Renewable Energy-Based Microgrids 217

Contents

        Creative Approaches to Project Financing 219
            Cap and Trade Programs 221
            Renewable Energy Certificates 223
            Global Environmental Facility and Clean Development
            Mechanism 223
            Build-Transfer Agreements 224
            Peer-to-Peer Trading 225
        Ways to Combine Project Financing and Delivery 225
            Public–Private Partnerships 226
            Energy Savings Performance Contracts 229
        Challenges to Microgrid Development 230
            Policies in Flux 231
            Development and Scalability Issues 231
            Controller Technologies 232
            Electricity Pricing 232
            Cost of Generation Systems, Integration, and Maintenance 233
        Cash Flow Analysis for a Microgrid Project 235
        Summary 239
        References 241

Chapter 15 It's Back to the Future with Microgrids 245

        It's All About Providing Electricity 245
            Bright Future for Microgrids 246
            Creating Smart Microgrids 246
            Development of Distributed Energy Resources 248
        Emerging Electrical Generation Technologies 249
            Wave Energy Systems 249
            Tidal Power 251
            Stirling-Dish Engine 252
            Electricity Generated Using Hydrogen 253
            Small-Scale Nuclear Reactors 254
            Plasma-Arc Gasification 256
        Developing Applications for Microgrids 257
            Artificial Intelligence 257
            Wireless Energy Transmission 258
        Conclusion 259
        References 259

Index 263

Preface

It seems like a long time ago when I first became interested in how we as a civilization might begin the process of weening ourselves off fossil fuels. Moving forward required an agenda that necessitated development on many fronts. This personal journey has been about energy conservation, energy efficiency, improved technologies, sustainability, and environmental improvements. It has focused on the types of policies we pursue and infrastructure we develop.

For me, solutions are found in congruent policies, programs, and technologies. Policies are established by governments and organizations. They set the goals and establish the tone of the agenda. Programs reinforce the policies by creating a supportive infrastructure that enables the policies to be communicated, funded, and subsidized. When the technologies are advanced and deployed, exciting things begin to happen. They are the creative endeavors of mankind that solve problems and provide needed services.

When I first started to understand the literature on the subject of microgrids, it seemed that most articles were too basic to be useful and most books were so detailed that only electrical engineering professionals could comprehend them. The authors often missed the policy aspects of microgrid deployment and failed to address the diversity of modern microgrids and their topologies. Microgrids were not being addressed as a new paradigm in the delivery of electricity. It gave me the idea that there was a need for a book about microgrids that could be understand by those who weren't necessarily engineers but would provide the fundamentals for those who were. It was this essential concept that lead me to pursue this project.

Fundamentals of Microgrids: Development and Implementation is about the structure of successful technological systems and how microgrids offer solutions for the future of local electrical generation. Microgrids in my view are the ultimate manifestation of the ideal of energy independence. They offer electricity to all takers, especially those with hopes of using renewables to generate local power, not totally dependent on the grid. They provide a multitude of potential solutions with the opportunity to do more with fewer resources. They bring electricity and its benefits to those in remote places and create stronger communities. The beauty lies in their adaptability and diversity.

Microgrids suggest that perhaps we already have the tools available to impact and resolve some of the key energy and environmental issues we face. If so, it is a matter of resolving where and how to deploy them. This creates lots of exciting opportunities for designers and engineers. To aid in this process, there are models and templates that we can learn from. This book is about these models and how we can organize the technologies available to provide infrastructure solutions that will become tomorrow's energy supply mechanisms.

Acknowledgments

During the course of my career, I have had opportunities to learn and share experiences with people who helped me understand how important it is to offer solutions to problems. In this endeavor, I have been humbled by what I could learn by studying the works of those who knew more than me about certain topics, such as microgrids. It is difficult to count the number of people who have helped in this pursuit, but many are found in this book's chapter references.

I must thank Al Thumann, Barney Capehart, Victor Ottaviano, Bill Payne, who all have passed on, for their insights, support, and friendship. I miss them all. Others who have helped in the past are Wayne Turner, Eric Woodruff, Bill Kent, Samer Zaweydeh, Fred Hauber, Michelle Whitlock, Joy Maugans, John Gilderbloom, and Steven Parker, who all have supported or assisted with my journal articles and books. I also thank the folks at Taylor & Francis Group, LLC who supported this project since the onset and believed that a book on this topic was not only timely but worthwhile.

My sincere thanks to all the contributing authors. Their contributions helped define areas of microgrid development that clarified situational applications. Without their aid this project would have lacked a well-rounded perspective. I also thank the students and professionals who have taken or supported my many seminars in renewable energy and microgrids over the last ten years or so. It was in those seminar sessions that many of my ideas about microgrids were initially presented, discussed, reevaluated, and finally congealed into terms that can be presented in a manner worthy of being published. I wish them well.

Editor Bio

Dr. Stephen A. Roosa, Ph.D., CEM, REP, BEP, CSDP, has over 35 years of experience in commercial energy management, energy engineering, and performance contracting. During his career he has been the corporate energy manager for a Fortune 100 company and has worked in various capacities with a number of energy services companies. His past experience includes thousands of energy studies and over $250 million in energy conservation, energy management, and performance contract projects that were developed for his customers.

Stephen is a former President of the Association of Energy Engineers (AEE). He currently serves as its Director of Sustainable State and Local Programs, and as the Chair of both the AEE Certified Sustainable Development Professional Board and the AEE Renewable Energy Professional Certification Board.

Stephen holds a Ph.D. from the University of Louisville, a Master of Business Administration from Webster University, and a Bachelor of Architecture from the University of Kentucky. He co-instructed his first graduate-level course at the University of Kentucky at the age of 19, implementing its first course in the use of the built environment as an educational medium. He has since taught graduate and undergraduate courses and professional certification courses and workshops throughout the U.S and in India, Saudi Arabia, Jordon, Kuwait, South Africa, and the Dominican Republic. Stephen is a Certified Energy Manager, a Certified Sustainable Development Professional, a Certified Building Commissioning Professional, a Certified Measurement and Verification Professional, a Certified Energy Monitoring and Control System Designer, a Certified Demand Side Management Specialist, a Certified Building Energy Management Professional, a Certified Renewable Energy Professional, and a LEED Accredited Professional.

Stephen is the Editor-in-Chief of the *International Journal of Strategic Energy and Environmental Planning*, a bi-monthly academic and professional journal. He is also on the Editorial Board of the *International Journal of Energy Management*. His published books include *International Solutions to Sustainable Energy, Policies and Applications* and *The Sustainable Development Handbook*. He was the Editor-in-Chief of the *Energy Management Handbook*, 9th edition. He has published over 50 journal articles relating to energy conservation, energy engineering, energy management, alternative energy, and sustainable development.

Stephen has received numerous related awards during his career including the AEE International Energy Manager of the Year Award, the U.S. Army Corps of Engineers National Energy Engineers Systems Technology Award, the U.S. Army Energy Conservation Award, and the U.S. Joint Chiefs of Staff Citation for Energy Management. He is a member of the Energy Manager's Hall of Fame and a two-time recipient of the Rockefeller Family Funds, Energy, and Environmental Education Award.

Contributors

Ishmael Ackah (Ph.D.) is an energy economist with experience in public service and academia. He has worked as the head of policy unit at the Africa Centre for Energy Policy, where he coordinated research on accountability and transparency in petroleum revenue management. He is currently the local content officer at Ghana's Energy Commission.

Alain Aoun received a Master of Science (2006) in industrial and power engineering and Masters of Engineering in electrical engineering (2007) and renewable energies (2018). He is presently the managing director of Alain Aoun and Partners, an engineering firm located in Lebanon that specializes in lighting, electrical, and energy systems. Alain Aoun is currently accredited with seven certifications from the AEE (CEM, BEP, CBCP, CEA, REP, CMVP, CLEP) and is a certified energy manager trainer. He is currently pursuing a Ph.D. in energy management.

Eric Banye is the sector leader (agriculture) and national project coordinator for the Voice for Change Partnership Program at the SNV Development Organization, Ghana. He has 14 years of experience in project management and project development.

Dramani Bukari is a senior energy advisor at the SNV Development Organization. He is a Ph.D. candidate at the Kwame Nkrumah University of Science and Technology. Dramani has facilitated major seminars including the World Bank's clean cooking seminar. He is an energy analyst with expertise in energy policy, energy modeling, small energy systems, project management, and economics.

Csaba Farkas was born in Berettyóújfalu, Hungary, on September 5, 1987. He received his Ph.D. from the Budapest University of Technology and Economics, Budapest in 2017. He is a senior lecturer at the Department of Electric Power Engineering and a network modeling engineer at the Hungarian Transmission System Operator MAVIR. His research interests include the computer simulation of electric power systems.

József Kiss received his Master of Science in electrical engineering from the BUTE in 2011. He is an assistant lecturer with the BUTE Department of Electric Power Engineering. His interests include power quality, network losses, renewable energy integration, and smart grids.

Eric Kyem is a head of department (economist) with the Energy Commission, Ghana. He is also the administrator of the electricity market oversight panel of Ghana (EMOP), an institution mandated by law to supervise the administration and operation of the Ghana Wholesale Electricity Market. Eric has worked successfully in various capacities in various organizations in the United Kingdom. He has experience in energy sector reforms and regulation. Eric holds a Master of Science in economics from the University of Surrey, UK.

László Prikler received his Master of Science in electrical engineering from the Technical University of Budapest in 1986, where he is currently an academic staff member in the Department of Electric Power Engineering. Mr. Prikler is a member of IEEE Power and Energy and Industry Applications Societies and the Hungarian Electrotechnical Association. He has received the Chapter Regional Outstanding Engineer Award of IEEE, Region 8. He is an honorary member of International Conference on Power Systems Transients (IPST) Steering Committee and served as chair of the European EMTP-ATP Users Group Association. He has co-authored more than 60 technical papers and 40 research reports. His research and development competence and interests include insulation coordination, analysis, simulation and measurement of power system disturbances, modeling and simulation of distributed energy systems, and the grid impact of electric vehicle infrastructure.

Shafic Suleman (Ph.D., ERP) is a lecturer at the Institute for Oil and Gas Studies, University of Cape Coast, Ghana. Shafic specializes in energy, sustainable development, and energy risk management. He has been involved in teaching, research, and consultancy services in energy and other related areas.

Attila Talamon received his Ph.D. in 2015 with research focusing on the Hungarian possibilities of low-energy buildings. He received a Master of Science in mechanical engineering from BUTE. He joined the Student Association of Energy in 2007; he is currently a senior member. He is a member of several professional organizations. He holds certifications as an energy auditor and building energy specialist. He has been involved in several international scientific projects as a leading expert. Since 2009 he has been lecturing on subjects related to renewable resources and building energy at BUTE and the University of Debreceni Egyetem (DE). He is currently with Szent István University. He is also a research fellow with the Centre for Energy Research, Hungarian Academy of Sciences.

István Vokony received his Master of Science in electrical engineering and his Ph.D. from the Budapest University of Technology and Economics (BUTE) in 2007 and 2012, respectively. He is a senior lecturer with the BUTE Department of Electric Power Engineering. He is a former officer of the AEE Hungary student chapter. His interests include system stability, renewable energy integration, energy efficiency, and smart grids. He is the corresponding author for this article.

List of Acronyms

AC	Alternating Current
AI	Artificial Intelligence
AMI	Advanced Metering Infrastructure
BESS	Battery Energy Storage System
BEMS	Building Energy Management System
BMS	Battery Management System
BREEAM	Building Research Establishment Environmental Assessment Method
CAES	Compressed Air Storage
CCGT	Combined-Cycle Gas Turbine
CCS	Carbon Capture and Sequestration
CCX	Chicago Climate Exchange
CDM	Clean Development Mechanism
CEER	Council of European Energy Regulators
CER	Certified Emissions Reduction
CERTS	Consortium for Electric Reliability Technology Solutions
CFI	Carbon Financial Instruments
CHP	Combined Heat and Power
CIGRE	International Council on Large Electric Systems (Fr.)
CPS	Clean Peak Standard
CSP	Concentrating Solar Power
CUF	Capacity Utilization Factor
DC	Direct Current
DDC	Direct Digital Control(s)
DER	Distributed Energy Resources
DERMS	Distributed Energy Resources Management Systems
DESS	Distributed Energy Storage Systems
DG	Distributed Generation
DMG	Dynamic Microgrid
DoE	Department of Energy (U.S.)
DR	Demand Response
DSO	Distribution System Operator
DSOC	Dynamic Stochastic Optimal Control
EDA	Energy Daily Allowance
EES	Electricity Energy Storage
EMS	Energy Management System
EN	European Norms
EPS	Electric Power System
EPA	Environmental Protection Agency (U.S.)
ERCOT	Electric Reliability Council of Texas
ES	Energy Storage
ESCO	Energy Service Company

ESMAP	Energy Sector Management Assistance Programs
ESPC	Energy Savings Performance Contracts
ESS	Energy Storage Systems
EV	Electric Vehicle
FAST	Feeder Automation System Technologies
FERC	Federal Energy Regulatory Commission
FESS	Flywheel Energy Storage System
GEDAP	Ghana Energy Development Access Project
GEF	Global Environmental Facility
GHG	Greenhouse Gases
GdV	Gorona del Viento
GOCO	Government-Owned Contractor-Operated
HES	Hybrid Energy Systems
HFC	Hydrofluorocarbon
HOMER	Hybrid Optimization of Multiple Energy Resources (software tool)
HP	Horse Power
HRES	Hybrid Renewable Energy Systems
ICE	Internal Combustion Engine
IEC	International Electrotechnical Commission
IECC	International Energy Conservation Code
IED	Intelligent Electronic Devices
IEEE	Institute of Electrical and Electronics Engineers
IRENA	International Renewable Energy Agency
ISO	Independent System Operator
IT	Information Technology
JIT	Just in Time
kW	Kilowatt
LED	Light Emitting Diode
LEED	Leadership in Energy and Environmental Design
LFG	Landfill Gas
LFGTE	Landfill Gas-to-Energy
LI	Lithium Ion
LICMP	Long Island Community Microgrid Project LICMP
LV	Low Voltage
MEMS	Microgrid Energy Management System
MW	Megawatt
MMTCE	Million Metric Tons of Carbon Equivalent
MPPT	Maximum Power Point Tracking
MSW	Municipal Solid Waste
MV	Medium Voltage
MW	Megawatt
M&V	Measurement and Verification
NARUC	National Association of Regulatory Utility Commissioners
NaS	Sodium and Sulfur
NG	Natural Gas

List of Acronyms

NRECA	National Rural Electric Cooperative Association
NREL	National Renewable Energy Laboratory
N-R HES	Nuclear-Renewable Hybrid Energy System
OECD	Organization for Economic Cooperation and Development
OFGEM	Office of Gas and Electricity Markets
OSG	On-Site Generation
O&M	Operation and Maintenance
OTEC	Ocean Thermal Energy Conversion
P2P	Peer-to-Peer
PCC	Point of Common Coupling
PESD	Portable Energy Storage Devices
POI	Point of Interconnect(ion)
PPA	Power Purchase Agreement
PURPA	Public Utility Regulatory Policies Act (U.S.)
PV	Photovoltaic
QECBs	Qualified Energy Conservation Bonds
RAPS	Remote Area Power Supply
RDEG	Renewable Distributed Energy Generation
RE	Renewable Energy
REC	Renewable Energy Certificate
RET	Renewable Energy Technology
RGGI	Regional Greenhouse Gas Initiative
RMS	Root Mean Square
RNG	Renewable Natural Gas
RPS	Renewable Portfolio Standard
SAPS	Stand-Alone Power System
SCADA	Supervisory Control and Data Acquisition
SDG&E	San Diego Gas and Electric
SEM	Subjective Evaluation Method
SGIP	Smart Grid Interoperability Panel
SHS	Solar Home Systems
SEPA	Smart Energy Power Alliance
SPS	Stationary Power Systems
TANGEDCO	Tamil Nadu Generation and Distribution Corporation
THD	Total Harmonic Distortion
TOU	Time of Use
TRC	Tradable Renewable Certificate
TS	Total Solids
TVA	Tennessee Valley Authority
UCSD	University of California at San Diego
UNT	Uniform National Tariff
UPS	Uninterruptible Power Supply
USC	University of the Sunshine Coast USC
US-CERT	United States—Computer Emergency Readiness Team
USD	United States Dollar

VPP	Virtual Power Plant
VRB	Vanadium Redox Batteries
VRE	Variable Renewable Energy
WFP	World Food Program (a UN initiative)
WTG	Wind Turbine Generators
ZNE	Zero Net Energy

Introduction

The fourth industrial revolution is associated with the global trend toward decentralizing energy grids. Within this context, microgrids are seen as a solution to how renewable electricity can be provided to local areas. *Fundamentals of Microgrids: Development and Implementation* provides an in-depth examination of alternative microgrid energy sources, applications, technologies, and policies. This book considers the fundamental configurations and applications for microgrids. It is a research-based work that discusses the use of microgrids as a means of meeting international sustainability goals. It focuses on questions and issues associated with microgrid topologies, development, and implementation. Technologies are discussed that are relevant to microgrid components and electrical generation systems. It considers both renewable and non-renewable generation systems as they apply to microgrids. Key themes include microgrid technologies, drivers for microgrids, types of microgrids, their components, configurations, and applications. Below is a summary of chapter content.

Chapter 1 provides a brief introduction to microgrids. It provides answers to the fundamental questions energy managers and other professionals want to know about the basics of microgrids. What are microgrids? When and why are they needed? They are often seen as solutions to improve electrical system resiliency and disaster response, and to reduce costs. How is this accomplished? This chapter begins by discussing the differences between conventional electrical power delivery systems and microgrids. The chapter provides definitions of microgrids and identifies the basic types of microgrids and the fundamental differences in microgrid types. Distributed energy resources are defined, standalone generation systems are described, and a typical microgrid configuration is provided. It describes what microgrids are and details the basics of how they work. An overview of the benefits of microgrids and their disadvantages is offered.

Chapter 2 offers an assessment of environmental drivers for microgrids. Our patterns of energy use since the early 1900s have contributed to the environmental pollution that is problematic today. With the arrival of the Hydrocarbon Age, fossil fuel consumption is increasing. How we use energy, and the types of energy we use, are both key to efforts to mitigate the environmental impacts of fossil fuel use. Like large-scale electrical grids, microgrids use fossil fuel-fired generation, renewable generation, and often both.

Developing sustainable policies requires foresight. Are microgrids a solution for some of the environmental issues associated with fossil fuel combustion? This chapter considers greenhouse gas-induced climate change as one environmental driver for using renewable energy resources in microgrids. The potential and costs of reducing greenhouse gases, while difficult to estimate, are considered. The economics of developing coal-fired plants is becoming more and more unfavorable. It seems more coal-fired thermal plants are being demolished or adapted for modified use than are being constructed. Options are being considered to enable a transition to a low-carbon economy. Examples include reducing use of targeted fuels, improving plant and transmission system efficiencies, and fuel substitution. The transition to a low-carbon economy is congruent with the development of microgrids. Also explored are additional environmental problems that include the emissions of criteria pollutants which reduce regional air and water quality. This chapter concludes with an optimistic forecast that microgrids have a place in solving key environmental issues.

Chapter 3 considers the roots and early history of microgrids. How did microgrids originate and evolve? This chapter traces the history of microgrids beginning with Thomas Edison's first microgrids in the U.S. Early microgrids were developed using both hydropower and fossil fuels. By the late 1880s, there were over 200 hydropower-driven microgrids in the U.S. Combined heat and power systems were co-located to provide electricity and heat to communities and factories. One example is the 1883 South Exposition in Louisville, Kentucky, which demonstrated the largest to-date installation of incandescent light bulbs invented by Thomas Edison and the first use of electric trollies. Westinghouse built an alternating current hydroelectric power plant on Niagara Falls and transmitted electricity 20 miles to Buffalo, New York. Combined heat and power systems grew to developing company and factory towns in the early 1900s.

This chapter describes models of microgrids that originated from marine applications and buildings. Ships are microgrids. Buildings that power themselves are considered to be nanogrids, a smaller version of a microgrid. Models for microgrids include marine examples such as military and cruise ships, residential nanogrids, plus green and net zero energy buildings. From these humble roots, microgrids have the potential to revolutionize our thinking about how buildings function and to disrupt the electric utility industry. Providing alternatives to fossil fuels for transportation systems is challenging. Today, the electrification of vehicular transportation systems provides justification for contemporary examples of microgrids.

Chapter 4 discusses traditional electrical energy systems and the development of large nation-scale generation systems. The electric grids of today consist of generating stations that produce electrical power, transmission lines that carry power to demand centers, and distribution lines that connect individual customers. These systems might use a wide array of generation including large hydropower, conventional fossil fuel systems, nuclear power, and in some cases renewable energy.

Central power stations typically rely on a single type of fuel for generation. Large central systems supply electricity power to consumers—a one-way model of dispatching power with inherent limitations. Developing problems with the electric grid include security issues, power disruptions, inefficiencies, environmental issues, and regulatory concerns. These problems have contributed to a complex system of electric regulation by federal, regional, and local governments to manage policies for generation, distribution, and sale of power to consumers. The regulatory environment often creates multiple complications and hurdles for distributed energy resources when interconnecting with the existing power systems and national grids. Overcoming these issues leads to successful microgrid development.

Chapter 5 evaluates microgrid architecture. It begins by describing microgrid operational configurations. How are grid-interconnected microgrids different from those that are not interconnected? This chapter details the key components of off-grid and grid-connected microgrids including centralized, decentralized, and distributed systems. There are several basic components of microgrid architecture. Most importantly, microgrids must contain the equipment necessary to generate electricity, loads or uses for the power generated, other assets, plus a microgrid power management system. Their topologies often include some form of energy storage system which does not necessarily need to be internal to the microgrid. They may be configured to allow the host electric grid, if interconnected, to serve as a storage system. Within the microgrid there must be a set of electricity-consuming devices that dictate the loads placed on the microgrid. Finally, grid-connected microgrids require a utility interconnection to enable the microgrid to exchange power with the larger utility network. Microgrids connect to large grids at the point of common coupling. Seen as a benefit of microgrids, their decentralized configurations have greater resiliency than centralized systems since their more diverse portfolio of production options provide generation redundancy.

Microgrid architecture configurations include models for various types of AC and DC electrical power generation and local distribution. Advanced microgrids are a subcategory of microgrids with defined characteristics, some similar to conventional microgrids but with enhanced capabilities. There are ways that microgrids can interconnect with other microgrids to improve operations and resiliency. Microgrids create a special regulatory issue as many states and local governments have not yet adopted regulations for their design and implementation. Some have no legal definition for a microgrid. However, many have regulations concerning the development of distributed energy resources and guidelines concerning utility interconnections. This chapter identifies and describes important microgrid regulations and standards.

Chapter 6 links microgrids to renewable energy solutions as distribution-scale technology options. How are renewables linked to microgrids? Alternative and renewable energies refer to electricity or heat generated from renewable sources plus energy substitutes and various conservation methodologies. Renewable energy resources have important policy, environmental, and economic impacts. They can be developed to effectively reduce greenhouse gas emissions in a sustainable manner and this is often cited as a driver for microgrid applications. Technology options for distributed generation include small hydropower, biomass energy, landfill gas extraction, geothermal energy, wind power, hydrogen fuel cells, solar thermal, and photovoltaics. Each is explored in the context of microgrid applications. All can be used to provide electrical generation for microgrids. The costs of renewable generation have been declining which opens new markets and provides increasing opportunities for deployment.

Chapter 7 considers the technical aspects of electrical energy storage systems. The storage of electricity creates interesting problems since electricity must be used the instant it is generated and cannot be directly stored. Technologies for storage change electricity to another form of energy and then regenerate power. What types of energy storage work best for microgrids? Examples of mechanical storage systems include pumped hydro, flywheel, and compressed air storage technologies. Electrochemical systems often used for microgrids encompass lead-acid, lithium-ion, sodium-sulfur, vanadium-redox, and zinc-air batteries. Other types of energy storage include chemical storage, biological storage, and thermal energy storage. In particular, molten salt storage systems are found in solar thermal microgrid applications.

All forms of electrical energy storage are limited by their capabilities, capacities, storage durations, and round-trip efficiencies. The technical and market barriers associated with distributed electrical storage are considered in this chapter along with ways for resolving these barriers. Storage systems can help microgrids become more responsive to utility demand programs and apply time-of-use rates to reduce electrical peak demand costs.

Chapter 8 rigorously examines hybrid energy generation technologies. How is hybrid generation defined? Hybrid generation systems from the perspective of microgrids involve the use of multiple generation and storage systems. The generation technologies used in microgrids have shortcomings which must be addressed to successfully apply microgrids. An important feature of microgrids are the mix of energy resources used in their designs for electrical generation. Some microgrids use multiple electricity generation technologies such as combinations of renewable fuels and fossil fuels. The idea is to find creative solutions to overcome shortcomings inherent in the use of any single fuel, thus meeting the goal of providing stable and resilient power.

There are innumerable combinations of generation systems that are possible. A few examples include combining diesel with renewable energy, combining solar thermal or photovoltaic with wind power, and using fuel cells or nuclear energy with renewables. Renewable co-generation and trigeneration systems are available for microgrid deployment. Storage technologies are often included as a key component of hybrid systems.

Chapter 9 considers local and regional microgrids. What causes communities and local organizations to consider a microgrid? Drivers include the desire for an independent power supply, the need to lower electricity costs, or the desire to improve resiliency. The ability to introduce renewable energy generation to meet sustainability goals is also cited. This chapter discusses community microgrids, campus microgrids, industrial microgrids, and military microgrids. It provides examples and case studies along with real-world microgrid applications. Case examples of microgrids that are considered include those serving Kodiak Island in Alaska, Borrego Springs in California, and Long Island in New York.

Virtual power plants, a special class of microgrids, are defined and discussed. They typically aggregate supply side resources, and often use a diverse pool of renewable distributed energy generation and wholesale renewable energy sources.

Chapter 10 considers small-scale rural mini-grids in developing countries as a solution for communities that are sparsely populated and located far from grid connections. These challenges have led to increased interest in off-grid and mini-grid solutions. Countries such as Ghana are not an exception to this trend.

How are microgrids successfully deployed in developing countries? This chapter highlights the viability of solar PV mini-grids for rural electrification in Ghana by analyzing the regulatory and fiscal situation. It offers recommendations for a supportive renewable energy and mini-grid regulatory framework. To these ends, this chapter provides a case study that examines the mini-grids deployed by the government and private investors and assesses the differences in tariff structures, customer services, and reliability. The findings indicated that expenditures on electricity using mini-grids were lower compared to the alternative of using kerosene and dry cell batteries. It was also determined that solar photovoltaic (PV) mini-grids developed by private businesses offer viable solutions for rural communities. Examples of the socioeconomic benefits of providing solar-generated electricity to rural communities include savings on fuel, improvements in the welfare of women, and educational benefits such as increased hours of learning. For rural areas fiscal incentives are needed to seed private sector investments in mini-grids. This chapter is authored by Ishmael Ackah, Eric Banye, Dramani Bukari, Eric Kyem, and Shafic Suleman.

Chapter 11 discusses microgrid islanding. There are locations (e.g., on remote islands, in mountainous areas) where the only way to provide electrical power is to establish local energy systems. What are the characteristics of island microgrids and what makes them successful? Local island microgrid solutions can be competitive in cases where the synchronous network is available, but new approaches are needed for economic reasons. One catalyst for this new approach to microgrid deployment is the growing renewable generation industry.

This chapter provides an in-depth island microgrid case study which considers the possibilities of local energy supply systems. In this chapter, service quality is analyzed, construction and operational requirements are cited, and through on-site measurements, a renewable energy mix optimizer application is introduced. Based on the measurements and operational experience, a calculation methodology is established to compare the traditional supply solutions with islanding operations. Using the comparison of network development costs and islanding solution costs, an optimal weight factor can be defined, which can be an effective investment decision support parameter. This chapter is authored by István Vokony, József Kiss, Csaba Farkas, László Prikler, and Attila Talamon.

Chapter 12 presents a detailed assessment of the use of an energy blockchain as a solution for developing energy resources. Blockchain offers a system in which a record of cryptocurrency transactions can be maintained across several computers that are linked in a peer-to-peer network. This can be applied to the transfer of funds for the purchase of electricity generated by microgrids. Considering the blockchain potential in developing countries from the distributed energy perspective, this chapter identifies the challenges and barriers associated with the development of DERs in developing countries and highlights opportunities associated with the integration of blockchain-based solutions to overcome these challenges.

In a future in which consumers of energy are encouraged to become producers, energy blockchains have an important role in the development of microgrids and smart grid frameworks. Blockchain has the potential to change the way transactions are processed. By eliminating the roles of third parties in transactions, blockchain can make our systems more economically efficient by reducing cost and improving reliability. The ability to conduct smart contracts for microgrid via blockchain platforms is shown to enable and expedite peer-to-peer energy trades which will further decentralize the energy markets of the future. Microgrid electricity sales and electric vehicle battery charging offer examples. Alain Aoun is this chapter's author.

Chapter 13 provides a detailed assessment of smart microgrids. What value is added by deploying smart technologies? As microgrids begin to morph into smart microgrids, new concepts have emerged that redefine how to deploy microgrids and incorporate new technologies. Information about energy production and loads is collected at a granular scale and algorithms analyze the data to effect efficiency improvements. Smart microgrids create additional benefits and expand opportunities for services by overlaying information age communication technologies. An objective of this chapter is to consider how smart systems are integrated into microgrids and enhance the transfer of electricity to and from the customer.

Smart microgrids use sustainable technologies and can be remotely controlled and managed. This chapter explains what smart microgrids are, how they work, and the key differences between traditional and smart microgrids. Smart microgrids often integrate suitable renewable energy technologies in an efficient manner to provide clean energy-based, direct, or alternating current microgrids. They are considered to be a revolutionary power solution. These types of microgrids are variously called advanced microgrids or advanced remote microgrids. They apply remote sensing, advanced metering, and digital communications. The chapter also provides real-world examples of smart microgrid projects under development and those already deployed. While there are concerns with policies that apply to advanced microgrids, the market for smart utility systems is expanding.

Chapter 14 provides an introduction into the key considerations that must be addressed to develop a business case for microgrid development. The interest in microgrids has been increasing as distributed energy resources have become more widely understood. The cost for renewables has declined making them more economically competitive. There are a number of ways to support the financing of microgrid projects. Traditional approaches include third-party financing, public–private partnerships, and build-transfer agreements. Microgrid development can be supported by sales of electricity or services, cap-and-trade programs, renewable energy certificates, clean development mechanisms, and peer-to-peer trading arrangements. Energy savings performance contracts can provide mechanisms for microgrid development.

Many microgrid developments are custom one-off projects that defy scalability and require custom engineering expertise. Some include costs for utility interconnections and electrical energy storage. Larger microgrids such as those that connect multiple centers and cross public rights-of-way face multiple policy barriers. Other challenges include non-standard controller technologies, variable electricity pricing schedules, and the costs associated with maintaining multiple generation systems. Financial risks must be identified during project development. Making the business case for microgrids necessitates an assessment of the value proposition, quantification of lifecycle costs and benefits, and the development of a detailed cash flow analysis.

Chapter 15 explores the future of microgrids. How are microgrids evolving and what does the future portend? For microgrids, it is all about electricity and the growing needs for the services it provides. There is a bright future for microgrids given the advances in distributed energy resources. Smart grids are emerging. The evolution of the electrical grid to incorporate monitoring, protect and automatically optimize the operation of its interconnected elements, and collect digital data instantaneously to respond automatically to variable conditions is already happening.

This chapter introduces nascent electrical generation technologies that may pave the way for future microgrid development. Examples include ocean wave and tidal energy, Stirling engines combined with solar reflectors, small-scale nuclear reactors, hydrogen generation, plasma-arc gasification, and others. These technologies have the potential to transform microgrids by creating a wider range of capabilities for generation technologies. Future microgrids will use artificial intelligence, provide control capability at a granular level for consumers, and transfer electricity wirelessly.

Microgrids provide a back-to-the-future solution for electrical energy systems. They provide opportunities to develop new networks targeted for the needs of localities and communities. If thoughtfully designed and carefully deployed, they can provide more resilient energy supply systems with greater environmental benefits.

1 Introduction to Microgrids

Microgrids are an exciting way to provide electricity to serve local needs and solve supply problems. They offer new ways to provide reliable and resilient electrical power. With the expanding use of decentralized energy resources, the role of microgrids in power supply systems is increasing as more are being developed.

Microgrids often consist of a number of small power supply systems which makes them more flexible than a single electrical power source. They can generate electricity from fossil fuels and renewable energy resources. They are categorized by their size, the types of customers they serve, the types of generation systems they have, and the regions in which they operate. They can be configured to produce either direct current, alternating current, or both. Some are semiautonomous and provide both heat and power. Their purposes vary from those of conventional electrical plants.

The categories of the energy we use for electrical generation are divided unevenly into nonrenewable sources; carbon-based energy sources such as coal, oil, and oil shale; and renewable sources such as wind power, solar, geothermal, and gravitational water sources. A conventional power station, also referred to as a power plant, powerhouse, generating station, or generating plant, is an industrial facility for the generation of electric power. Most power stations contain one or more generators, rotating machines that convert mechanical power into electrical power. Conventional power stations typically use fossil fuel-fired generators, most notably coal, natural gas, and nuclear power, using the Rankin cycle. However, there are many others that use renewable technologies such as hydroelectric dams and large-scale solar power stations. Such stations are centralized and require electric energy to be transmitted over long distances [1].

Renewable energy sources can be categorized as sustainable and inexhaustible, while most nonrenewable energy sources are potentially unsustainable and likely exhaustible. Microgrids are alternatives to conventional power stations for generating power. Technologies are converging that enable microgrids to be seen as a new and viable solution to providing locally generated electrical power. They are the next step in the evolution of supplying and delivering electricity.

WHAT ARE MICROGRIDS?

Today, large power plants commonly use coal, gas, hydroelectric, and nuclear energy. They are centralized. Electricity is transmitted over long distances. Like central power stations, microgrids generate electricity. Being small-scale versions of the larger electrical grids, they are sometimes referred to as mini-grids. Microgrids have become more common with the growth of renewable electrical generation. While microgrids share common characteristics with utility grids, they are not identical.

The electricity generated by microgrids is primarily consumed by local users within the boundaries of the microgrid. They are often custom configured to solve local electrical generation problems and may use combined heat and power systems.

As there does not seem to be a universal definition as to what a microgrid is, a number of various definitions are offered. According the U.S. Department of Energy, a *microgrid* is a "group of interconnected loads and distributed energy resources within clearly defined electrical boundaries that acts as a single controllable entity with respect to the grid" [2]. A microgrid can connect and disconnect from the grid to enable it to operate in either grid-connected or island modes. When thus configured, seamless transfer of electric power is important for system operation. The International Council on Large Electric Systems (CIGRR, Fr.) C6.22 Working Group defines *microgrids* as "electricity distribution systems containing loads and distributed energy resources (such as distributed generators, storage devices, or controllable loads) that can be operated in a controlled, coordinated way either while connected to the main power network or while islanded" [2]. Its definition is qualified by the presence and type of electrical generator, storage devices (electrical, pressure, gravitational, flywheel, and heat storage technologies), and controlled loads [2]. Examples of controlled electrical loads include automatically dimmable lighting or pumping systems. Yet another definition: "Microgrids are modern, small-scale versions of the centralized electricity system. They achieve specific local goals, such as reliability, carbon emission reduction, diversification of energy sources, and cost reduction, established by the community being served. Like the bulk power grid, smart microgrids generate, distribute, and regulate the flow of electricity to consumers, but do so locally" [3].

The term *microgrid* today refers to a collection of multiple smaller energy resources of different types (including renewables such as solar arrays and wind turbines but also natural gas generators, combined heat and power systems, and energy storage) often owned by local *prosumers*, small businesses or small power operators interconnected to supply the locality with self-sustaining power [4]. Microgrids located in the U.S. and other Organization for Economic Co-operation and Development (OECD) countries have capacities in the range of hundreds of kilowatts (kW) or megawatts (MW) [4]. Utility-scale microgrids in the U.S. typically exceed 10 MW.

In many non-OECD countries, microgrids of this scale (sometimes called *minigrids*) are the primary electrical power source for people in rural communities where access to the central grid is either too costly or unreliable, or is perhaps nonexistent [4]. The label *microgrid* in countries such as India is reserved for systems under 10 MW. Though India is striving for universal access to electrical power, other parts of the world are struggling. With 13% of the world's population, Africa accounts for only 4% of the world's energy demand, leaving roughly 600 million people with no access to electricity [5].

STANDALONE POWER

Examples of technology options for electrical supply can be roughly viewed as two scales, the nature of which varies: 1) *utility-scale*—large, bulk power applications with better economics but experienced indirectly through purchases of premium

Introduction to Microgrids 3

electricity products with some renewable energy content; 2) *distributed-scale*—small, on-site applications experienced directly but somewhat limited due to economics and resource availability yet having greater opportunity for renewables [6]. Modern standalone electrical generation systems are usually considered to be distributed-scale applications.

A *standalone power system* (SAPS or SPS), also known as a remote area power supply (RAPS), is an off-the-grid, independent electricity system for locations that lack an extensive utility-scale electricity distribution system. The main categories of technological components for a SPS are: 1) components that supply power (generation equipment); 2) components that store energy for later use (energy storage equipment); and 3) components that convert one form of power to another (inverters) and those that control the flow of power in a system (power conditioning and control equipment) [7]. An SPS is often used to provide electricity for remote buildings, telecom sites, villages, or small-scale manufacturing (e.g., mining, resource processing) sites. In such situations, connecting to the electric grid via rural electrification programs is likely more expensive than installing an SPS. Since power is provided only to a local geographical area, transmission costs are minimal. Examples of an SPS include diesel or natural gas generators and solar photovoltaic systems. For owners, the most common reason to have an SPS is that a grid connection is not available. Having an SPS might also appeal to their desire for independent power, better align with their environmental values, or reduce their operation and fuel costs associated with generating electricity.

Typically, standalone power systems include one or more methods of electricity generation, energy storage, and regulation. They can be configured to act or be used as an uninterruptable power supply. They often take advantage of a combination of components and technologies to generate reliable power, reduce costs, and minimize inconvenience [8]. On-site generation (OSG), district/decentralized energy, or distributed generation systems generate or store electricity from a variety of small, grid-connected devices called a distributed energy resource (DER) or distributed energy resource systems (DERS) [1]. Some of the strategies for SPS that deploy distributed energy resources (DERs) include reducing the amount of electricity required for loads (via energy efficiency or load shedding) and using fossil fuel or renewable hybrid systems [8]. Many remote SPS that previously relied on diesel fuel are being augmented or replaced with renewable energy systems as prices for diesel fuel increase or become more volatile.

DISTRIBUTED ENERGY RESOURCES

Microgrids are designed to satisfy the demands of energy consumers within their limited service areas. Their local low- and medium-voltage electrical distribution systems contain a set of connected electrical loads and generation from DERS. Electrical switch gear, either manually or by automatic controls, is used to isolate the microgrid at the point of common coupling (PCC). For automatic or standby mode, the system operates as a single-utility service in the event of a power failure [9]. It can also operate in automatic peak-shaving mode for paralleling with the utility to reduce consumption of the power it supplies [9]. Manual-mode operation provides a wide latitude of control when conditions warrant using the system as a supervised station [9].

DER systems for microgrid applications may use fossil fuel or renewable energy sources, or combinations of both. Examples include thermoelectric generation, small hydro, micro combined heat and power (CHP), diesel, biomass, biogas, solar power, wind power, and geothermal power that connect to existing power distribution systems. DERs allow for infrastructure development in ways that protect against singular events and vulnerabilities [10]. Grid-connected devices for electricity storage can be classified as DER systems and are often called distributed energy storage systems (DESS). By means of an interface, DERs can be managed and coordinated within a smart grid. Distributed generation and storage enable the collection of energy from many sources while lowering environmental impacts and improving security of supply [1].

Within defined boundaries, microgrids provide a controllable electrical supply which maintains equilibrium between supply and loads. Many microgrids can connect and disconnect from the electric grid. This allows them to operate in either *island* (standalone, possibly emergency state) or *grid-connected* (normal state) modes. They normally have equipment that generates electricity and low-voltage transmission, plus a power management system. They are linked to energy storage systems and electricity consuming devices. Some are semiautonomous. The International Renewable Energy Agency (IRENA) says that microgrids (by OECD standards) are installed "to achieve exceptionally high levels of reliability for industrial applications, such as data farms or industrial processes for which a power outage could prove extremely costly" (Figure 1.1) [4].

Why Would You Need a Microgrid?

It is surprising that despite the growth of the world's electrical grids, there are over a billion people on the planet that lack access to electricity, a prime mover of the information age. Many have no opportunity to be connected to an electrical grid. Lacking power for lighting, devices, and electronics, they have little use for the appliances people in the developed world use on a daily basis. Without grid access, how can people become more educated, use cell phones, learn to use computers, or access the internet? Alternative ways to provide electricity must be considered.

The electric grid connects buildings to central power sources, which allow appliances, heating/cooling systems, and electronics to be used [11]. The need for electrical power is the fundamental reason to have grid-connected power which is why microgrids are required. Those without access to grid-connected electricity might want to have a source of electrical power, possibly an electric generator. If the generator failed, a backup source (a different type of generation or storage) of some kind would be needed. However, a suitable and reliable system could be powered by microgrids using distributed generators, renewable resources, and energy storage.

The key reason to have a microgrid is to generate electricity reliably and locally. Unless an adequate value proposition in favor of developing a microgrid is provided, the arguments for having microgrids are weakened. If a locality has access to grid-supplied electricity, then electrical power that uses diversified fuel sources that are

Introduction to Microgrids 5

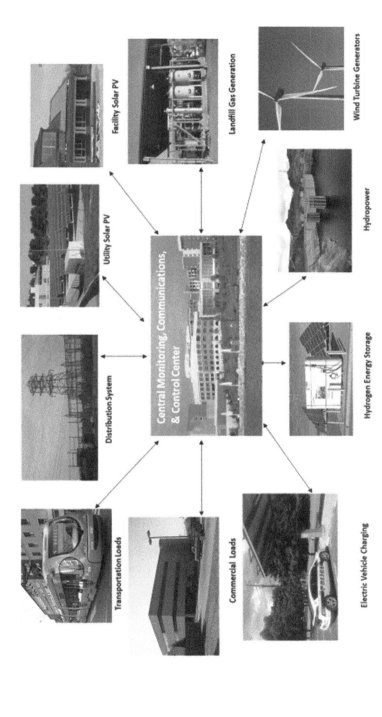

FIGURE 1.1 Microgrid configuration with central monitoring and control.

less costly, more dependable, more resilient, more secure, less troublesome, more environmentally appropriate, and/or have fewer associated externalities is one solution. If none of this applies to a given situation, there may be no need for a microgrid.

The interconnectedness of the central electric grids means that when part of the grid needs to be upgraded, everyone on the grid is potentially impacted [11]. This is especially true when high-voltage distribution and transmission infrastructure must be repaired or replaced. A primary purpose of a microgrid is to have available, adequate electricity for non-interruptible and critical loads [12]. Microgrids connect to the grid at a point of common coupling. A microgrid generally operates while connected to the grid, but can disconnect and operate on its own using local energy generation in times of crisis such as major storms or power outages [11]. This scenario assumes that the microgrid is unaffected by the disturbance that causes grid power to be interrupted. When not connected to the main grid, a microgrid operates as an independent entity.

Mitigating grid disturbances and improving grid resilience are often mentioned as goals for microgrid development. For power systems, *resiliency* means more than returning to an original state. It refers to the ability to harden from or recover quickly from a high-impact, low-frequency event, improving the ability to persist when compromised [13]. Ideally, microgrids can operate reliably and continuously when properly managed using alternative energy generation equipment and storage.

Types of Microgrids

Microgrids usually consist of a number of small power supply systems. Some produce either direct current (DC) or alternating current (AC). While AC microgrids are the norm, a DC microgrid may be used to improve reliability and efficiencies, enhance operational performance, or provide advanced capabilities that enable control of network resources for the ability to operate independently of the primary AC system [14]. Some microgrid plants are configured to provide both heat and power. Microgrids can be configured to be interconnected with the electric grid (see Figure 1.2a) or independent of the grid (see Figure 1.2b).

There are two major types of microgrids—*true microgrids* and *milligrids*. Microgrids wholly on a single site, akin to a traditional utility customer, are usually called customer microgrids or true microgrids ($\mu grids$); those microgrids that involve a segment of the legacy regulated grid are often called milligrids (*mgrids*) [15]. Milligrids allow distributed energy resources to be deployed so they can be directed to critical infrastructure in the event of emergencies. Another type is remote power systems (*rgrids*), which are isolated and unable to operate in a grid-connected mode [15].

Microgrids are generally categorized by their features and applications. They can be either off-grid or grid-interconnected. They can be categorized by their use by utilities, industries, commercial entities, institutions, universities, communities, and military bases. The microgrid at Fort Bliss in El Paso, Texas, which is grid-connected, would be classified as a grid-interconnected military microgrid.

Microgrids are a proven way to provide power to remote areas. In the Canary Islands, a microgrid supplies electricity using wind, solar photovoltaic (PV), and

Introduction to Microgrids

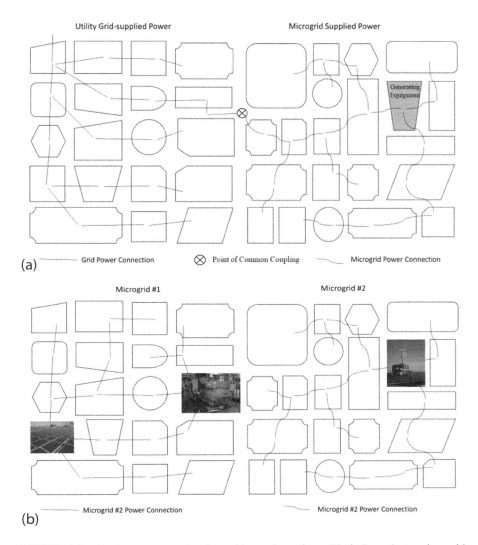

FIGURE 1.2 (a) Interconnected microgrid configuration; (b) independent microgrid configuration.

pumped hydro storage. The Caribbean island of Bonaire has a microgrid. Power is supplied by wind and biodiesel generation and supplemented with a battery storage system. Both of these are examples of island microgrids that are off-grid systems.

Cosidine et al. provide an interesting alternative classification for types of microgrids [16]. They include common categories such as military and industrial microgrids but further identify others: 1) *development microgrids*, those for small commercial operations with existing infrastructure; 2) *motivational microgrids*, those that operate in the absence of compelling needs; 3) *hidden microgrids*, those with generators on-site that they use to provide emergency power; and 4) *isolated microgrids*, those for isolated vacation homes in the mountains or on islands, or perhaps for yachts [16]. For military and emergency power response in the event of disasters, *mobile microgrids* are a solution. These are often used for remote, strategic applications when electricity is immediately required and there is no short-term potential to connect to grid-supplied power sources. Many of these use diesel generation or solar power for electrical generation and can be operated remotely if needed. They may have difficulty matching loads with generation and may require some type of microgrid control system.

Within the last few years, there have been efforts to distinguish between microgrids as commonly defined and other categories of microgrids. Similar to many microgrids, an advanced microgrid is capable of automatically interacting with, connecting to, and disconnecting from another grid [16]. *Advanced microgrids* contain all the essential elements of a large-scale grid, are dynamic, and have the ability to: 1) balance electrical demand with sources, 2) schedule the dispatch of resources, and 3) preserve grid reliability (both adequacy and security) [17]. They grant users the flexibility to securely manage the reliability and resiliency of the microgrid and connected loads while mitigating the economic impacts associated with power disruptions [17]. A key feature of advanced microgrids is the ability to achieve plug-and-play interoperability within the sphere of the technologies used for electrical generation and compatible communication.

Some microgrids can be classified as *virtual microgrids* (*vgrids*). These include DERs located at multiple nonadjacent sites that are coordinated so that they can be presented to the grid as a single entity but operate virtually as a controlled island or coordinated multiple islands [15]. Virtual microgrids are often loose aggregations of individual generation sources and loads that can be remotely controlled. In this case they use the infrastructure of the host grid and while unable to be decoupled physically, they are operated within the energy market as if independent [18]. They can be configured based on software connectivity (cloud-based).

How Does a Microgrid Help?

The electric grid connects buildings to central power sources, which allow the use of appliances, heating/cooling systems, and electronics. This interconnectedness means that when part of the grid needs to be repaired, everyone is potentially affected [19]. This is how a microgrid helps. It connects to the grid at a point of common coupling which maintains voltage at the same level as the main grid unless there is a problem on the grid or other reason to disconnect. A microgrid generally operates while connected to the grid, but can disconnect and operate on its own using local energy generation when required [19]. Microgrid internal configurations provide service versatility. They can provide electricity to a single user, serve a partial or full feeder, or provide service at a full or distribution substation (see Figure 1.3).

Microgrids can provide power to remote areas that lack the option of connecting to a grid. In such cases, it operates as a standalone utility system providing electricity to its connected loads. Controls manage output based on the demand of the customer services. With multiple generation systems, the least costly and most reliable energy

Introduction to Microgrids

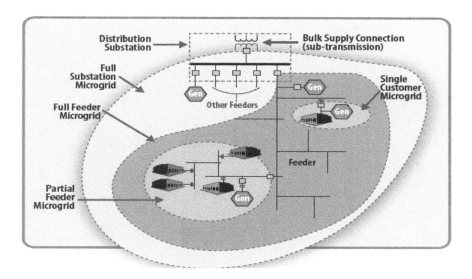

FIGURE 1.3 Typical microgrid configurations. (Source: U.S. Department of Energy [20].)

resources can be utilized. Depending on the types of fuels required and how its load requirements are managed, a microgrid can be designed to operate indefinitely [19].

ADVANTAGES AND DISADVANTAGES OF MICROGRIDS

Microgrids generate electricity from a diversified range of electrical generation sources and can use a variety of technologies. When renewables are used for generation processes, microgrids emit less pollution and have lower carbon footprints. There are other advantages and disadvantages associated with the deployment of microgrids.

Central grids are costly to maintain and subject to disruption. Improving electrical transmission resiliency in the U.S. is costly and often hindered by state regulations. The U.S. is experiencing more weather-related electrical outages caused by winds, flooding, and ice storms. These occur in many states, including California due to wildfires, the Gulf states due to hurricanes, and the Midwest and Northeast states due to more intense winter conditions [21]. Recurring flooding has occurred along the Mississippi River basin and in the Great Lakes. A wildfire in California sparked by electrical transmission wires killed 85 people in 2018 and caused $16.5 billion in damages [21]. Ice storms are notorious for taking down electrical power lines, causing millions of dollars in damages and requiring long periods of time to repair under stressful conditions. Researchers have estimated that there were 679 widespread weather-related electrical system outages in the ten-year period ending in 2012 [21]. Excluding major events U.S. electric customers in 2017 experienced an average of 1.4 power interruptions with 7.8 hours of total outage (see Figure 1.4) [22]. The states with the most total hours of outages were Maine, Florida, New Hampshire, Georgia, and Vermont with Maine having 42 hours [22].

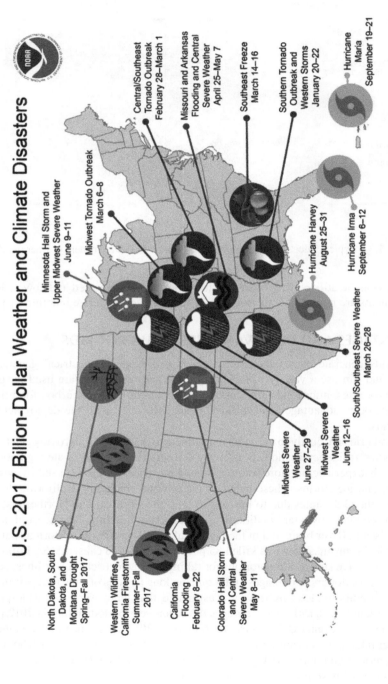

FIGURE 1.4 U.S. climate disasters in that occurred in 2017. (Source: U.S. National Atmospheric and Oceanic Administration.)

Much of the weather-related damage to electricity distribution systems in the U.S. could be prevented if overhead power lines were not used. While greater transmission resiliency can be achieved by placing transmission underground, cost is a substantial burden—a single mile of underground transmission can cost as much as $3 million to install and widespread deployments would substantially increase electrical costs [21]. As a result, less than a quarter of new electrical cable is buried and the rest consists of overhead transmission lines mounted on utility poles which are exposed to weather and falling tree limbs.

ADVANTAGES OF MICROGRIDS

An important benefit of microgrids is the potential to improve electrical system reliability. Microgrids often have multiple power generation sources and many deploy both fossil fuels and renewables in an effort to improve reliability and absorb intermittency. DERs enable electrical power to be generated at or near the loads which substantially reduces transmission costs and improves resiliency. This makes them more flexible and resilient than a single electrical power source. Less transmission infrastructure is needed as energy is generated locally. Since line voltages are much lower repair costs are less.

Another advantage is that many microgrids provide dispatchable power to critical loads. This provides electrical power that is available upon demand. Generators can be turned on or off, or can have their power output adjusted as needed. Hydropower, biomass, tidal power, and geothermal energy can be designed to be dispatchable without energy storage. Examples of renewable energy sources normally referred to as non-dispatchable include solar and wind power. These DERS allow for energy infrastructure deployment that protects against single vulnerabilities [10]. Power may be also stored in batteries or reservoirs to provide dispatchable generation of electricity.

Microgrids provide customers with alternative power generation. A compelling and fundamental feature of microgrids is their ability to separate and operate in isolation (known as islanding) from the utility's distribution system during blackouts [23]. Islanding occurs when a microgrid or segment of the utility electrical grid with both loads and distributed generation is isolated from the rest of the grid yet continues to operate independently from the power grid [24].

Intentional islanding is the act of physically disconnecting a set of electric circuits from a utility system, and operating those circuits independently [25]. When this occurs, the island is seen by the utility grid as a single electrical load. These islands must be supplied from suitable sources and able to guarantee acceptable voltage support, frequency, controllability, and quality [24]. Non-intentional islanding occurs after a fault when it is impossible to disconnect the distributed generation system. It is imperative that non-intentional islanding be detected and quickly eliminated [24].

To comply with grid requirements distributed generation systems must automatically shut down during times of power outages [23]. This makes microgrids appealing to entities that experience high costs from electrical outages even if they are of relatively short duration. Continuous industrial processes and emergency lighting systems are common examples. Microgrids can be intentionally operated as islands during power outages that occur during extreme weather events. From an electrical quality perspective, microgrids can offer power-smoothing, shift electrical generation and loads, and provide seamless power transfer.

Climate scientists believe that severe weather events are becoming more common due to the impacts of climate change. Since microgrids are smaller distributed generation systems, they offer redundancy during system failures and provide targeted power delivery to address a locale's specific requirements [10]. Microgrids typically have black-start capability since multiple generation resources within the microgrid allow the system to restart on its own [26]. Black start is the process of restoring power to part of an electric grid without relying on the external electric power transmission networks. Interestingly, some utility companies are installing microgrids for their central headquarters to enable them to operate as command centers to coordinate response activities during massive area-wide outages [27].

There is economic justification for microgrids and they can reduce costs in a number of ways. In some cases, developing a local microgrid can reduce or avoid the capital costs needed to expand the macrogrid to accommodate increased load, peak power requirements, or power quality issues. For example, when a feeder or substation upgrade is required to address increased loads or power quality, a microgrid with on-site generation could satisfy the need without a significant capital investment [28]. When multiple generation sources are available, decisions can be made as to which type of fuel source is the least expensive at a given time. Microgrid management systems can be designed with the logistics to reduce costs by incorporating advanced peak-shaving capabilities and to profitably arbitrage energy pricing differences. Algorithms can be used to reduce risks and selectively energize loads during operations and extended outages. Since generation sources are normally in close proximity to loads, local generation reduces the costs of transmission infrastructure and the associated losses inherent in transmitting electricity over long distances. When renewable generation is included in the microgrid, the business risks associated with variable fossil fuel costs is reduced or eliminated.

Revenue can be generated by selling excess power to the grid if the microgrid is interconnected. In some cases, it is possible to provide wider services to the grid and obtain payments supported by feed-in tariffs [29]. In the U.S., microgrid owners have opportunities to participate in state and federal clean energy programs (e.g., renewable portfolio standard initiatives), federal production tax credits, and grant programs that specifically target microgrid development [27].

Disadvantages of Microgrids

There are disadvantages to using microgrids. Development and maintenance costs can be expensive. This is particularly true when they have multiple electrical generation systems. A solar plant might be coupled with a wind power site to provide continuous electricity. If the microgrid relies on grid power, there are added costs for interconnection equipment. A storage system (e.g., batteries, compressed air, or pumped storage hydropower) may be required to ensure uninterruptable electrical power.

Economics and customer preferences are causing developers to integrate greater amounts of non-dispatchable renewables (50% to 100% of capacity) into their microgrids [28]. The use of some types of renewable energy (e.g., solar and wind power) can present intermittency problems and system-balancing challenges which must be addressed. This is especially true if there is an over-reliance on intermittent generation. If the microgrid has low capacity, there may be a highly dynamic load situation to address. While costs for microgrid development are much less than supporting a conventional utility grid, they provide much less power and can cost more on a per-kW basis. Finally, a fundamental issue is that the engineering expertise to develop and maintain the microgrid may not be readily available. If the system is very remote, the costs of maintenance and service at the remote location can be higher than anticipated.

Another disadvantage that thwarts microgrid development involves imposed limitations by policymakers. Often, regulations concerning microgrid development are unclear or nonexistent. The key challenge is that the multi-user microgrid often requires that the primary utility grant an exception to its established (typically regulated) utility franchise rules [12]. Incumbent electric utility companies often resist microgrid operations within their established service territories. Some utility executives are reluctant to embrace local renewable generation due to fears that the existing power grids are unable to reliably integrate distributed energy generation [30]. Such grid reliability concerns have effectively limited many local renewable projects to providing no more than 15% of peak power needs [30].

SUMMARY

Microgrids are a back-to-the-future solution. Electricity is a prime mover of the information age. Those without access to an electrical grid are literally left in the dark without power for lighting, appliances, and electronics. While countries such as the U.S., China, and Brazil have grid access rates over 95%, other countries are not as fortunate. While access to electricity is ubiquitous in most parts of the world, about 16% of the world's population (an estimated 1.2 billion people) still lack this basic necessity [31]. The continent of Africa is particularly in need of access to electrical power (see Figure 1.5). In such areas of the world, microgrids are seen as a potential low-cost solution to providing electricity. Access to electricity is a notable driver for microgrid development to support local needs. Others see microgrids as a solution to meeting environmental goals or obtaining power that is more reliable and perhaps less costly. Microgrids provide opportunities for access to a local electrical system.

Microgrids can be configured to be grid-interconnected or -independent. There are a number of different types of microgrids including customer microgrids, minigrids, virtual microgrids, and remote power systems. They are designed to satisfy the demands of the energy consumers within their limited service areas. Microgrids can be powered by distributed generators, batteries, and/or renewable resources. Examples of distributed energy resources that are used to power microgrids include thermoelectric generation, small hydro, micro combined heat and power, diesel, biomass, biogas, solar power, wind power, and geothermal power. There is a movement from diesel-fired generation toward a greater use of renewable energy resources to generate electricity.

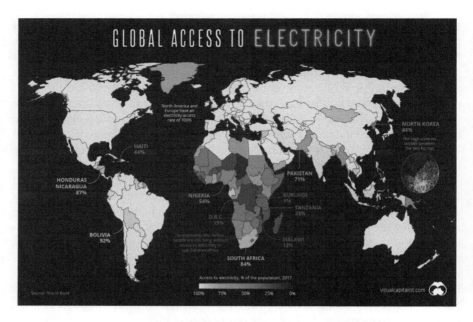

FIGURE 1.5 Global access to electricity. (Source: www.viualcapitalist.com [31].)

Many microgrids can connect or disconnect from the electric grid at will and offer the potential for distributive-scale, standalone power. They connect to existing power distribution systems at a point of common coupling. With the advantage of multiple generation systems, the least costly and most reliable energy resources can be deployed.

There are advantages and disadvantages associated with the deployment of microgrids. With access to grid-supplied electricity, a microgrid is desirable if cleaner electrical power is possible using diversified fuel sources. Among the advantages is the potential to improve transmission system resiliency with less investment. Other benefits of microgrids include reducing electricity costs, improved reliability, security, and reduced transmission infrastructure. Transmission system efficiency and resiliency can be better achieved with less investment when microgrids are deployed. Another compelling and fundamental feature of microgrids is their ability to separate and operate in island mode, disconnected from the host electric grid.

The disadvantages of microgrids include economic, technical, and policy challenges. Microgrids require engineering design and custom deployment. Economic issues include financing difficulties and the potential of higher costs for development and maintenance. If microgrids are not connected to a local grid, energy storage may be required to ensure reliability. The greatest disadvantage that thwarts microgrid development involves imposed limitations by policymakers and incumbent industries. Multi-user microgrids often require that the primary utility grant an exception to its established utility franchise rules. Regardless, there is great potential for microgrids.

REFERENCES

1. Wikipedia (2018, October 17). Distributed generation. https://en.wikipedia.org/wiki/Distributed_generation, accessed 3 December 2019.
2. Microgrids at Berkeley Lab (2018). Microgrid definitions. https://building-microgrid.lbl.gov/microgrid-definitions, accessed 9 February 2020.
3. Galvin Electricity Initiative (2018). What are smart microgrids? http://www.galvinpower.org/microgrids, accessed 9 February 2020.
4. Coman, C. (2017, January 30). How the world defines microgrids and why you are confused. https://microgridknowledge.com/defines-microgrids, accessed 31 October 2018.
5. Wood, E. (2016, August 30). Microgrid roadmap building continues in California... ViZa to bring microgrids to Africa with Jabil Inala. https://microgridknowledge.com/microgrid-roadmap-california-viza, accessed 31 October 2018.
6. Roosa, S., editor (2018). *Energy Management Handbook*, 9th edition. Fairmont Press: Lilburn, Georgia, pages 453–460.
7. Murdoch University. Stand-alone power supply systems. http://www.see.murdoch.edu.au/resources/info/Applic/SPS, accessed 20 December 2018.
8. U.S. Department of Energy, Office of Energy Efficiency and Renewable Energy. Off-grid or stand-alone renewable energy systems. https://www.energy.gov/energysaver/grid-or-stand-alone-renewable-energy-systems, accessed 20 December 2018.
9. Porter, G. and Sciver, J., editors (1999). *Power Quality Solutions: Case Studies for Troubleshooters*. Fairmont Press: Lilburn, Georgia, page 38.
10. Lumbley, M. (2018, November). Beyond politics, a project manager's argument for clean energy. *North American Clean Energy*, 12(6), page 64.
11. Lantero, A. (2014, June 17). How microgrids work. https://www.energy.gov/articles/how-microgrids-work, accessed 31 October 2018.
12. KEMA (2014, February 3). Microgrids – benefits, models, barriers and suggested policy initiative for the Commonwealth of Massachusetts, pages 6–7.
13. Gellings, C. (2015). *Smart Grid Implementation and Planning*. Fairmont Press: Lilburn, Georgia, page 230.
14. Blackhaus, S., Swift, G., Chatzivasileiadis, S., Tschudi, W., Glover, S., Starke, M., Wang, J., Yue, M. and Hammerstrom, D. (2015, March). DC microgrids scoping study—estimate of technical and economic benefits. Los Alamos National Laboratory. https://www.energy.gov/sites/prod/files/2015/03/f20/DC_Microgrid_Scoping_Study_LosAlamos-Mar2015.pdf, accessed 16 January 2019.
15. Berkeley Lab (2018). Types of microgrids. https://building-microgrid.lbl.gov/types-microgrids, accessed 24 January 2020.
16. Cosidine, T., Cox, W. and Cazalet, T. (2012). Understanding microgrids as the essential architecture of smart energy. https://www.researchgate.net/publication/267927596_Understanding_Microgrids_as_the_Essential_Architecture_of_Smart_Energy, accessed 24 January 2020.
17. Sandia National Laboratories (2014, March). The advanced microgrid: integration and interoperability. Sandia Report SAND2014–1535.
18. LO3 Team (2018, February 16). All about microgrids. https://lo3energy.com/find-energy-industry-needs-microgrids-work-now-work-future, accessed 3 November 2018.
19. U.S. Department of Energy (2014, June 17). How microgrids work. www.energy.gov/articles/how-microgrids-work, accessed 5 February 2020.
20. U.S. Department of Energy. The role of microgrids in helping to advance the nation's energy system. https://www.energy.gov/oe/activities/technology-development/grid-modernization-and-smart-grid/role-microgrids-helping, accessed 23 January 2020.
21. Kicknyer, E. and Gecker, J. (2019, November 27). Above-ground power lines get riskier as nasty storms grow. *Courier-Journal*, page 19A.

22. Walton, R. (2018, December 4). Electric power outages in 2017 doubled in duration: EIA faults large storms. https://www.utilitydive.com/news/electric-power-outages-in-2017-doubled-in-duration-eia-faults-large-storms/543526, accessed 9 February 2020.
23. Asmus, O. (2009, November 6). The microgrid revolution. http://peterasmus.com/journal/2009/11/6/the-microgrid-revolution.html, accessed 3 November 2018.
24. Nabi, N. and Anusha, S. (2016, November 11). Voltage sag and support in power control on distributed generation inverters. *International Journal for Modern Trends in Science and Technology*, 2(1), page 194.
25. Microgrid Institute (2014). About microgrids. http://www.microgridinstitute.org/about-microgrids.html, accessed 11 November 2018.
26. S&C Electric Company (2018). Is a microgrid right for you? https://www.sandc.com/globalassets/sac-electric/documents/sharepoint/documents---all-documents/education-material-180-4504.pdf, accessed 24 January 2020.
27. Wood, E. (2018, November 4). Microgrid benefits: eight was a microgrid will improve your operation… and the world. https://microgridknowledge.com/microgrid-benefits-eight, accessed 5 December 2018.
28. GTM Research (2016, February). Integrating high levels of renewables into microgrids – opportunities, challenges and strategies. http://www.sustainablepowersystems.com/wp-content/uploads/2016/03/GTM-Whitepaper-Integrating-High-Levels-of-Renewables-into-Microgrids.pdf, accessed 11 January 2020.
29. Interesting Engineering (2015). 7 benefits of microgrids. https://interestingengineering.com/7-benefits-of-microgrids, accessed 11 February 2020.
30. Clean Coalition. Community microgrids. Bringing communities an unparalleled trifecta of economic, environmental and resilience benefits. http://clean-coalition.org/community-microgrids, accessed 24 January 2020.
31. Routley, N. (2019, November 2019). Mapped: 1.2 billion people without access to electricity. *Visual Capitalist*. https://www.visualcapitalist.com/mapped-billion-people-without-access-to-electricity, accessed 28 June 2020.

2 Environmental Drivers for Microgrid Development

INTRODUCTION

Hydrocarbon-based energy consumption is increasing. This is primarily due to emissions from developing countries. Carbon dioxide (CO_2) is released into the atmosphere by natural sources and the combustion of carbon-containing fuels and petroleum-based distillates. Atmospheric carbon and greenhouse gas (GHG) concentrations that drive global climate change have the potential to negate our boldest efforts toward mitigation.

How we use energy and the types of energy we use are keys to our mitigation efforts. Sustainable energy sources are an important component of our solutions. Renewable energy (RE) resources can be categorized as *sustainable*, while most nonrenewable energy sources are *potentially unsustainable*, and likely exhaustible. There is hope that we have time to reduce the introduction of GHGs into the atmosphere by implementing climate stabilization remedies. The increased deployment of RE in microgrid configurations is among the solutions often cited as being more sustainable.

The geophysical and biochemical processes that store carbon require hundreds of millions of years for the renewal process to be effective. Much of the solar energy captured and stored in the Earth by fossilized hydrocarbon processes that are being extracted and consumed today date from the Paleozoic period, roughly 600 million years ago—predating mankind's existence. At that time natural processes worked for millions of years to remove high concentrations of CO_2 from the atmosphere, storing them safely away in subsurface geologic formations. Most carbon-based energy being consumed as primary fuels are actually fossilized biomass and are not categorized as being renewable. These GHG-containing fuels include peat, lignite, oil, coal, and natural gas (e.g., methane). Petroleum distillates that generate carbon include gasoline, kerosene, propane, and diesel fuels. Fuels sources such as those derived from waste streams (e.g., wood, sawdust, landfill gas, animal wastes) are more difficult to classify as they are carbon-based fuels yet are rapidly renewable if the waste streams are maintained. These are typically classified as biomass and considered to be renewable fuel sources.

We understand the serious problems concerning GHG emissions and know that they are directly related to the combustion of fossil fuels. The emissions of atmospheric carbon generated by man's activities are exacerbating environmental changes on a global scale. The incremental increases in atmospheric carbon emissions have been successfully tracked; the importance of finding solutions that reduce CO_2 emissions cannot be understated. GHG concentrations in the atmosphere are increasing, trapping more heat. In the past, we seem to have made conscious choices to ignore the potential consequences. Today, the political and social structures we have established are developing embryonic policies toward solutions yet are taking only feeble actions. Our initiatives often seem mismatched, our actions marginal, and their impacts minimal. We have pushed aside our planet's stewardship responsibilities and failed to bring our technical and economic resources to bear on the task of implementing viable solutions.

Hope for solutions can be found in developing and implementing relevant policies, programs, and technologies. Governments are creating policies, corporations are reconstructing strategic plans, and institutions are redefining their missions. Agendas are in flux and new programs are being launched and implemented. There is an evolving consensus on the horizon, one that will change how we prioritize our efforts to become more sustainable. Innovative technologies, such as renewable energy systems and microgrids, are being proven and developed. They have the potential to transform energy supply systems to ones that are carbon-neutral. This is already happening at the local and regional scale in some areas of the world. RE deployment has a key role and microgrids provide opportunities to expand their deployment.

While the world is moving toward greater use of renewable energy, hydrocarbon-based energy consumption continues to increase. There are many obstacles in managing environmental problems. One of the most common is that many environmental issues are seen as international in scope rather than local. This perspective diminishes the impetus to attack many problems with local efforts. Successful response requires international treaties such as the Paris Agreement.

This chapter considers the issue of atmospheric carbon emissions and other environmental issues that are among the drivers for considering the deployment of microgrids. It discusses the importance of solutions that focus on renewable energy technologies as the key to long-term sustainability and the evolution toward a low-carbon economy.

THE HYDROCARBON AGE HAS ARRIVED

New environmental problems have surfaced that emphasize the ultimate fragility of our planet's ecosystems. Due to our unresponsive choices, climate changes are occurring at a rapidly accelerated pace. Our common future is becoming grim. Despite actions we might take today, the consequences of increasing carbon levels will continue to plague the Earth for generations to come. Never before have human-induced environmental changes had the potential for global impact. These environmental changes have been festering for a long time and local impacts from climate change can be devastating.

Our patterns of energy use since the early 1900s have contributed to the increasingly dire circumstances we find ourselves in today. The results of our past insensitivity to the environment are becoming more and more noticeable, the consequences less and less dismissible. With the acceptance that an over-abundance of atmospheric CO_2 is impacting Earth's ecosystems in detrimental ways, a new age is dawning. The economies of the Hydrocarbon Age are at an impasse. Indeed, the availability of certain fossil fuels is reaching a tipping point in some regions of the world. We are changing our landscapes as a result of economic development and technological advancements—terraforming them to meet the needs of agriculture, industry, urban expansion, and regional transportation networks. Many of our cities have evolved to become post-industrial and are unlike anything that our forbearers might have imagined. Cities in Pakistan, Brazil, Mexico, China, and India are reeling from explosive population growth, poor development choices, and environmental damage. Air pollution from fossil fuel combustion is making daily life difficult. Yet access to energy remains a key to sustainability. It enables mass education, access to domestic water supplies, the cultivation and distribution of food stocks, and the provision of health care.

Countries seeking economic and political control over carbon-based energy resources are at times exerting their military muscle, threatening and waging regional conflicts. For example, Iran attacked several oil tankers in the Persian Gulf and a Saudi Arabian refinery in 2019 in an effort to disrupt the distribution of crude oil and refined products. The impacts of each country's policies and economic systems are intensifying the problems associated with carbon emissions rather than resolving them. However, the political dialogue is shifting from outright rejection of the existence of carbon-induced climate problems to the development of strategies and legislation to mitigate them.

Sustainability requires foresight. At this point in history—when creative solutions are needed—the backdrop of conflicting influences, political stalemates, and economic turmoil makes long-term solutions difficult to implement. Energy is linked to the world economy and energy-consuming systems account for 95% of man-made CO_2 emissions (see Figure 2.1). Atmospheric carbon concentrations are approaching 410 ppm, the highest levels in the history of mankind. How can we cost-effectively reduce the energy-related emissions of GHGs to stabilize the climate while meeting the increasing demand for energy? The seminal inconvenient truth is that a consensus regarding how to proceed remains elusive. The high stakes are evident. All that is required is an increase in atmospheric temperatures of 2°C to 3°C and the climate will evolve to being like it was in the Pliocene period, about three million years ago, when sea levels were estimated to be about 25 meters (80 feet) higher than today. Already we are seeing ice shelves melt away, warmer seas, increased coastal flooding, and stronger storms. In late 2019, the city of Venice, Italy was inundated by the rising seas of the Adriatic, the worst flooding in over 50 years.

Mankind's continued misuse of carbon-based fuels has created uncertainty and increased risks. While there are both natural and man-made sources of atmospheric carbon, natural endowments that reduce atmospheric carbon levels are under attack by human activities. Formed over hundreds of millions of years, carbon compounds were stored by natural processes that worked very slowly. These processes gradually reduced atmospheric concentrations by storing carbon in natural sinks, such as soils, vegetation, below the Earth's surface, and the oceans. Over time, they transformed the Earth and created conditions to enable human evolution. There are currently no viable substitutes for these natural processes that have effectively sequestered massive quantities of carbon for eons.

Many of the adverse environmental consequences we face are caused by increased atmospheric carbon concentrations and their impact has only recently

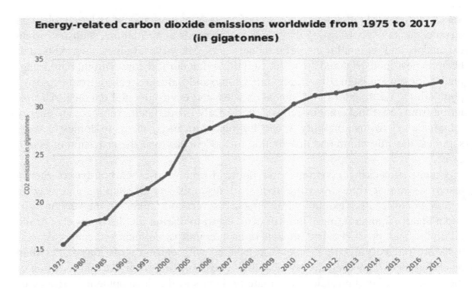

FIGURE 2.1 Global carbon dioxide emissions. (Source: IEA, Statista 2018.)

become understood. Why is this? As with serious environmental problems encountered in the past (e.g., natural resource depletion, species extinctions, or water pollution), there is a time lag before the environmental impacts of human actions become noticeable, measurable, and verifiable. Environmental changes must be meticulously observed before the causes and effects are assessed. Theories next evolve as to why the changes are occurring. Potential impacts must be weighed. To do this, the scientific community must make comparative measurements of before and after conditions, link the environmental changes to the causes, and consider an assortment of remedial actions. Prior to suggesting approaches to mitigation, a wide array of intervention methodologies and technologies must be developed, tested, and deployed. After a period of societal denial and disbelief, a political consensus must emerge before public resources can be used. This process is time-intensive and laborious.

A scientific consensus regarding the environmental damage caused by carbon emissions has only recently become available. Increasingly, the deleterious consequences of carbon-induced climate change have become more apparent. The consensus is that carbon emissions are contributing to possibly irreversible changes in the world climate. The time of inaction has passed and the time of initiative is upon us. We must take action now else our common future is in peril.

Carbon Dioxide Emissions

Today, the carbon compounds that are being increasingly released into the atmosphere come from many sources, including primary combustion of fuels such as coal, natural gas, and oil. Atmospheric concentrations of CO_2 are increasing. The consumption of fossil fuels is the primary cause of both CO_2 emissions and the discharge of pollutants such as sulfur dioxide, nitrogen oxides, mercury, and fine particulates [1]. Ways must be implemented to reduce carbon emissions and the pollution generated from the use of fossil fuels.

A consensus has evolved that global climate change is a result. We know that carbon emissions are directly linked to the use of fossil fuels and they are responsible for global climate change. Mankind's activities are exacerbating this problem and creating climatic disruption, loss of biodiversity, and economic uncertainty. Despite increased awareness, efforts to address the impacts of climate change remain in their infancy. They have yielded little impact on global emissions. In fact, carbon emissions worldwide continue to increase, reaching 37.1 billion metric tons of CO_2 in 2018. This is due to the growing quantities of fossil fuels being extracted and how they are used in combustion processes.

Global carbon dioxide emissions (CO_2e) from man-made sources more than tripled from 1950 to 2000 and they continue to increase. From 1990 to 1999, global emissions increased at a rate of 1.1% annually, jumping to 3% annually despite the creation of the Kyoto Protocol [2]. Projections indicate that CO_2e will increase to 42.9 billion metric tons by 2030 [3]. Increases of such magnitude are unprecedented. The potential impacts on the Earth's ecosystems are unknown. It is likely that further damage to Earth's climate will occur. This damage will have unforeseen consequences and impact our economies, our health, our resources, and our settlements. These changes are likely to occur so quickly that we may be unable to adjust rapidly enough to avoid global catastrophe.

Another potent GHG, methane (CH_4), is at least 25 times more damaging than CO_2 when initially released. Methane gas is used as a combustion fuel for electrical production and escapes into the atmosphere from fermentation, landfills, coal mining operations, manure, natural gas systems, and other sources. Since methane combustion releases less CO_2 per unit of heat generated than other hydrocarbon fuels, it is seldom the fuel of choice. With a half-life of seven years, atmospheric methane can oxidize producing CO_2 and water [4]. In comparison to CO_2, each methane molecule has a relatively large global warming impact that diminishes in a comparatively short period of time. Methane tends to concentrate in the stratosphere and in tropical regions. For these reasons, efforts to mitigate carbon concentrations must include preventing the release of methane gas into the atmosphere. The amount of methane present in the atmosphere has increased from 700 parts per billion (ppb) in 1750 to 1,745 ppb in 1998, and stands at 1,866 ppb in 2019 with more than half of the emissions caused by human activity [4].

There is wide variability in national carbon emissions rates and emissions are unstable and unevenly distributed [5]. Atmospheric carbon emissions would likely have risen faster were it not for the 7% decline among industrial countries since 2007—a group that includes the United States, Canada, Europe, Russia, Australia, New Zealand, and Japan [6]. Atmospheric carbon emissions in the U.S. in 2018 were estimated to be 5.27 gigatons. In 2018, ten countries were responsible for two-thirds of CO_2 equivalent emissions.

In more populous, developing countries, the rates of emissions increase are becoming unmanageable. In 2004, China and India combined for 22% of world emissions, yet their share is anticipated to further increase to 31% by 2030 [3]. In 2010 the Chinese government enacted a series of polices requiring methane from coal mining to be captured, or converted into CO_2, yet the new policies failed to curb total emissions [7]. China surpassed the U.S. in 2016 as the world's No. 1 carbon dioxide polluter. The largest portion of China's anthropogenic GHG emissions are attributable to coal mining [7]. Despite efforts to reduce GHG emissions in the U.S., as of 2018, the U.S. had the dubious distinction of being among those countries with the highest emissions of GHGs gases on a per capita basis and was responsible for about 15% of all GHGs emissions. The U.S. emits CO_2 at more than twice the per capita rate of China (see Figure 2.2).

Within countries, carbon dioxide emissions vary widely across economic sectors (see Table 2.1). GHGs from power generation, industrial processes, and transportation account for over 52% of the world's total greenhouse gas emissions. CO_2 emissions are often concentrated in or near urban areas and vary widely. In 2014, Canadian cities produce roughly 15.2 tons of CO_2 per capita per year, while residents of the Netherlands produce only 9.9 tons annually on a per capita basis [9].

INTERNATIONAL POLICIES—KYOTO AND THE PARIS AGREEMENT

One of the biggest obstacles in creating policies to reduce atmospheric carbon emissions is that many environmental issues are seen as international in scope rather than local. It is often the sum of local actions which is most effective. Existing institutions and national bureaucratic structures often cannot handle and were not created to deal with such problems effectively. Without the institutional capacity in place necessary to handle problems on an international scale, inertia and inaction are often the result. While local and regional responses can have powerful impacts, supra-national

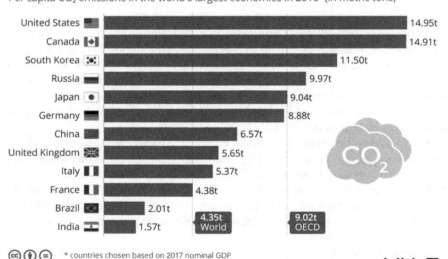

FIGURE 2.2 Per capita 2016 carbon emissions in top emitting countries. (Source: Statista [5].)

TABLE 2.1
Total 2018 Carbon Dioxide Emissions (Adapted from [8])

Country	CO_2 emissions (Billion metric tons)	Share of total (%)	Δ after Kyoto (% change)
China	9.43	27.8	+54.6
U.S.	5.15	15.2	−12.1
India	2.48	7.3	+105.8
Russia	1.55	4.6	+5.7
Japan	1.15	3.4	−0.1
Germany	0.73	2.1	−11.7
South Korea	0.7	2.1	+34.1
Iran	0.66	1.9	+57.7
Saudi Arabia	0.57	1.7	+59.9
Canada	0.55	1.6	+1.6
Total	22.97	67.7	

governmental organizations (e.g., the United Nations and the European Union) have also proved successful. For example, international action to prevent additional damage to the ozone layer required cooperative and concerted efforts and proved to be effective, evolving into the Montreal Protocol.

Treaties such as the Kyoto Protocol and the Paris Agreement were designed to reduce greenhouse gasses. The Kyoto Protocol is an international treaty which became active in February 2005 had a goal of stabilizing GHG concentrations in the atmosphere at a level that balances anthropogenic (manmade) interference with the climate system. This 2015 pact committed countries to limiting global warming to 2° above pre-industrial level with 1.5° being the preferred limit [1].

Despite having not approved the Kyoto Treaty, arguing that the required decreases in emissions were impossible to achieve and efforts to do so would cause economic ruin, the U.S. has substantially decreased emissions since the Kyoto Protocol was enacted. Surprisingly, it has also met most of its emissions reduction requirements—one of the few countries to do so. In fact, U.S. CO_2 emissions in 2018 were only 8% greater than those reported in 1990 though the country's population grew from 248.9 million in 1990 to 327.9 million in 2018, a 32% increase.

The Paris Agreement, signed by representatives of 193 countries in April 2016, was a step toward GHG policy at an international scale. Laurent Fabius, France's foreign minister, said this agreement is an "ambitious and balanced" approach and a "historic turning point" in our goal of reducing global warming. Perhaps its most important goal is to quickly achieve the global peaking of GHG emissions. Many of the countries participating in this agreement have made substantial GHG emission reductions which illustrates the success of their progressive measures (e.g., Armenia, Azerbaijan, Hungary, Romania, Sweden, Switzerland, Kyrgyzstan, Georgia, Germany, Denmark, and others). Ultimately, we need to create a pathway for the Earth's climate resilience. However, for the Paris Agreement to be effective, all countries must cooperatively participate. As a result of administrative bias, the U.S. withdrew from the Paris Agreement in 2019 though it remains a party to the United Nations Framework Convention on Climate Change [1]. While the administrative branch of the U.S. federal government abstains, many state and local governments have established GHG reduction programs, upending federal leadership.

Carbon Management—Adapt or Mitigate?

From a technical perspective, there are a number of ways to manage atmospheric levels of carbon. Emissions can be reduced or carbon can be extracted from the atmosphere. Reducing emissions is the least costly and disruptive. Like other atmospheric pollutants, once emitted into the atmosphere, carbon is diffused and becomes difficult and expensive to extract. The technologies available to remove carbon from the atmosphere are commercially unproven, costly, and not scalable. Thus, reducing the quantities of carbon compounds prior to their release into the atmosphere holds far greater promise.

An alternative is to adapt to the new and developing circumstances. As the climate warms and sea levels rise, we can move inland seeking higher ground. Yet this is costly and requires the abandonment of many low-lying coastal communities and cities. Infrastructure losses will exceed hundreds of trillions of dollars. The world's cities at risk of major flooding and economic damage include Athens, Manila, Dhaka, Istanbul, Jakarta, Shanghai, and London. Those in the U.S. include New York, Houston, New Orleans, and Miami. This is already happening locally. A village in Alaska was the first in the U.S. to be moved inland by the Army Corps of Engineers, at a cost exceeding $250 million. The documented justification stated the cause at rising seas due to climate change. Other economic impacts have even greater costs.

When management and adaptation fail, mitigation often proves costly. The state of Florida has expended over $1.3 billion for beach restoration over the last 70 years using rising sea levels due to climate change as the justification. Another recent example is the ironic request of oil refining industries in southwest Texas. Their representatives have consistently denied the effects of climate change while their products have contributed to atmospheric carbon levels. Yet they have recently requested the use of state and federal taxpayer funds, several hundreds of millions of dollars, to build levies to protect them from increasing sea levels which they say are being caused by rising water levels from the Gulf of Mexico.

To complicate matters, there is no single solution to our carbon management problems. The *smoking gun* is often more readily identified than the elusive *silver bullet*—and none may exist. In fact, scientific investigations often conclude that a set of concerted mitigation approaches are necessary. Such contextual responses vary in their identification of the source of the problems, local circumstances, economic costs, and scope of the solution. In some cases, deploying recommended technologies has unintended results that are not necessarily favorable. In the case of microgrids that rely on wind turbines, the disposal of used rotor blades can create problems for local landfills. In the face of mounting evidence to the contrary, some believe that carbon management is unnecessary as the Earth's environment is self-correcting. While this is arguable for the long-term (thousands of years), the near-term (hundreds of years) impacts to our environment remain unaddressed. The potential costs of future mitigation increase to the point of being unimaginable.

Our patterns of energy use will dictate the success of our mitigation efforts. Sustainable energy sources are an important component of our solutions. The keys to actually reducing carbon emissions include reducing the use of carbon-based fuels by improving efficiencies and fuel substitution. The concentrations of carbon in fossil fuels vary, as do the amounts of carbon released into the atmosphere during combustion processes. New energy sources must produce fewer carbon emissions. One way to do this is to use renewable forms of energy. These are forms of energy that are derived from and replaced rapidly by natural processes. They include solar thermal and photovoltaic, wind power, hydropower, tidal power, biomass, and geothermal energy.

COSTS OF REDUCING GREENHOUSE GAS EMISSIONS

The cost of reducing U.S. greenhouse gases is difficult to estimate. The projections range from almost nothing to hundreds of billions, if not trillions, of dollars. According to a report entitled *Reducing U.S. Greenhouse Emissions: How Much at What Cost?* by McKinsey and Company, annual GHG emissions in the U.S. are projected to increase to 9.7 gigatons in 2030 if no remedial actions are undertaken [10]. This report identifies the primary causes for the projected growth of U.S. carbon emissions:

- the anticipated long-term expansion of the economy
- growth in the use of energy by buildings, appliances, and transportation due to a projected population growth increase of 70 million
- the continued reliance on carbon-based electrical power generation from the construction of new coal-fired power plants that lack carbon capture and storage (CCS) technology
- a gradual decline in the ability of U.S. forests and agricultural lands to absorb carbon, forecasted to decrease from 1.1 gigatons in 2005 to 1.0 gigatons in 2030 [10]

The McKinsey and Company report provides case projections that consider abatement opportunities such as greater use of coal with CCS and expanded use of nuclear power, renewable energy, and biofuels, along with vehicle efficiency improvements and energy efficiency upgrades for buildings. While the use of coal is declining in developed countries such as the U.S. and Germany, it is increasing in China and India [8]. Most carbon abatement projections establish a cost of $50 per ton or less. While such costs might incentivize projects for fossil fuel-fired thermal plants, motor vehicle efficiency improvements typically require greater levels of investment to reduce emissions.

In the U.S., policy focus on climate change is widely divergent. GHG mitigation costs are often cited as the reason for inaction though the costs are not identified. While the federal government lacks focus and centers its efforts on partisan funding for local remediation projects that lack long-term impacts, many states and local governments are moving forward with more effective mitigation policies. The U.S. Climate Alliance, representing 17 states and territories, has decided to observe the country's responsibilities under the Paris Agreement, supplanting federal leadership [1]. Many cities and states are committing to being more reliant on renewable energy rather than fossil fuels, citing renewables as being less costly.

POTENTIAL FOR REDUCING CARBON EMISSIONS

There remain opportunities to further reduce carbon emissions. The potential in the U.S. can be measured by how policies, programs, and technologies are effectively synchronized to implement this goal. There must be a focus on managing broadly based initiatives that successfully reduce carbon emissions. However, the pressures of population growth and development temper the belief that carbon emissions can be reduced by only initiating additional energy conservation and efficiency improvements. Technological solutions, such as pre- and post-combustion carbon capture are expensive and untested at large scales. Finding terrestrial carbon sinks can be difficult since storage of carbon for long periods of time (hundreds or thousands of years) is not guaranteed.

The study by McKinsey and Company evaluated five categories of technologies that can be employed to reduce carbon emissions and estimates their potential impacts [10]. They are:

- reducing the carbon intensity from electric power production by using alternative energy and CCS technologies (800–1,570 megatons)
- improved energy efficiency in buildings and appliances (710–870 megatons)
- implementing carbon reduction opportunities in the industrial sector (620–770 megatons)
- expanding natural carbon sinks to capture and store more carbon (440–590 megatons)
- increasing vehicular efficiency and using less carbon-intensive fuels (340–660 megatons) [10]

These approaches offer an estimated total reduction in carbon emissions ranging from 2,910 to 4,460 megatons, or a reduction from the base year (2005) emissions ranging from 40% to 62%. Substantial energy efficiency infrastructure improvements will be necessary to meet such goals. If implemented in concert, these initiatives would place the U.S. at a per capita carbon emission level approximating that of Germany of Japan.

The U.S. is improving the efficiency of vehicles and new buildings. Yet it is clear that the country needs to direct more resources and efforts toward reducing emissions from electrical energy production and improving the efficiency of existing buildings and appliances—areas where the greatest reductions in carbon emissions are possible. The need to expand the U.S. economy and support a growing population makes the implementation of carbon emission mitigation programs challenging. However, there are ways to reduce the impact of carbon emissions without adversely impacting the economy. Reducing electrical demand (kW) is a prime example. Projections indicate that improvements in building and appliance energy efficiency combined with industrial sector initiatives could offset 85% of the incremental demand for electricity through 2030 [11]. Microgrids designed to boost power to the grid during peaking conditions offer a potential solution. This scenario is certainly possible, yet the increased need for electricity through 2030 must be met while decreasing the number of coal-fired plants.

Renewables provide us with lower, scalable CO_2 emissions options for energy production [12]. Nuclear power generation is a potentially viable option as it does not generate CO_2 emissions. However, the high development costs, long planning horizons, and nuclear waste disposal issues associated with nuclear power stations remain largely unresolved. Until these concerns are addressed nuclear energy is not considered to be a universally viable solution.

THE NEW ECONOMICS OF COAL PLANTS

The use of coal represents a large portion of the CO_2 being released into the atmosphere. Coal use is decreasing in the U.S.; in 2018, coal use declined to levels not seen since 1979. Yet it is increasing in some developing countries. Despite its high levels of CO_2 emissions and its low thermal energy conversion efficiency (only about 37%), coal usage has an important role in global energy supply.

The electric utility industry's plans to decrease coal-fired electrical power generation is a result of economic realities and U.S. energy and environmental policies. In 2009, the AES Corporation withdrew an application to construct a 650 MW coal-fired power plant near Shady Point, Oklahoma as part of a strategy to reevaluate their growth plans [13]. Louisiana Generating was recently sued by the Environmental Protection Agency (EPA) for failing to install modern pollution control equipment when its Big Cajun Plant underwent major modifications [14]. Michigan has placed several power plants on hold while Governor Jennifer Granholm attempts to shift her state to power from cleaner, more sustainable energy sources [14]. In Kansas, a state with a successful wind power development program, former Governor Kathleen Sebelius vetoed three legislative attempts to approve two large coal plants at Holcomb with a total generating capacity of 1,400 MW as was initially proposed by Sunflower Electric Power Corporation [15]. In the U.S. the largest GHG declines have surprisingly come from lower emissions from coal (primarily used to generate electricity) as power generation switched to less expensive natural gas and the use of carbon-free wind energy [6].

Perhaps one of the strongest arguments against the development of coal generation concerns its long-term economics. The startling fact is that coal-fired power plants are fast becoming non-competitive. Investors are becoming resistant to financing coal plants and insurers are becoming hesitant to underwrite their potential liabilities. Most states are using less coal. In Connecticut, New Hampshire, and Massachusetts, coal consumption has declined by over 75%. In others, coal has effectively become an intermittent power source. The Energy Information Agency determined that natural gas-fired power plants are less expensive to construct with overnight capital costs ranging from $676 to $2,095 per KW, depending on the technology used [16]. This is the primary reason why natural gas now generates 34%

of the electricity in the U.S. The math shows that for coal to compete with natural gas, coal would require additional subsidies from $1.2 to $4.5 billion per coal plant constructed. Natural gas plants can also be started more quickly when needed and do not have ash disposal problems that require management.

From the microgrid developer's perspective, it is almost antidotal that substituting natural gas for coal reduces CO_2 emissions. This is due to the lower carbon content of natural gas and our ability to combust it at higher efficiencies than coal. Capacity additions have favored natural gas and renewable energy while most power plants being retired used coal [17]. Lower natural gas prices, increased regulation on air emissions, and more efficient natural gas-fired combined-cycle technology have made natural gas an attractive choice for baseload demand previously met by coal-fired generation [17]. Coal-fired generation has decreased because of economics.

What about conventional coal-fired thermal plants built to meet present standards without carbon capture? These wouldn't necessarily be called "dirty coal" plants but they do cause carbon pollution. Interestingly, they are often not price competitive either. They cost $2.9 to $6.6 million per MW to construct plus the power plant owners need to continue purchasing coal [16]. When compared to greener technologies, coal plants have permitting issues, are more costly to build, take longer to construct, and are more expensive to operate. New wind power plants cost about $1.3 to $2.2 million per MW and their costs are declining—plus the utilities don't need to purchase the fuel source. Concurrently, costs for solar photovoltaic and thermal plants continue to decline. A solar project in Ohio was recently estimated to cost $1.1 million per MW.

Facing such a competitive environment coal companies creatively find ways to maximize the subsidies from the government. To reduce state and federal royalty payments and maximize subsidies from the U.S. government, coal companies actually created subsidiaries to sell the coal they mine to themselves [18]. The scheme involves captive transactions in which companies that mine the coal use other companies they own to sell the resource at a low value. This reduces payments to the federal government since the first sale of the coal is at a below-market price to a subsidiary [18]. In 2017 coal producers asked the federal government to subsidize the development of coal-fired plants, arguing that they are more resilient than other forms of generation and are needed for energy security. Governmental subsidies make the cost of coal-fired generation appear to be economically feasible when it is not. Regardless, economic realities can be harsh. To reduce its energy costs, the Kentucky Coal Museum, located in a state that has historically supported the development of coal plants, installed solar photovoltaic panels on its roof. Coal use has been declining in the U.S. while natural consumption has been increasing (see Figure 2.3).

TRANSITIONING TO A LOW-CARBON ECONOMY

Strategies to reduce carbon emissions will impact all sectors of the economy and costs are likely to be unevenly distributed regardless of the strategies employed. Therefore, policies that focus on reducing carbon emissions must be broadly based. A consensus must be reached. Programs must be implemented to achieve the goals

Environmental Drivers for Microgrid Development

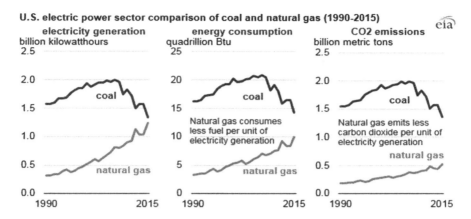

FIGURE 2.3 U.S. energy impact of reduced coal use. (Source: Energy Information Administration [17].)

established by the policies that are adopted. While policies and programs are replicable, they must contain overarching principles, have consistent goals, and be locally adaptable. The technologies employed must be appropriate, deployable, and economically viable. There are a number of strategies that would enable a transition to an economic model based on lower carbon emissions.

GREENHOUSE GAS REDUCTION STRATEGIES

A strategy that is widely used in environmental policy-setting is called stabilization triangle theory. For greenhouse gasses, the concept involves stabilizing emissions at a point in time causing them to initially level off and then decline over a period of years (see Figure 2.4). This is difficult as GHG emission declines must occur in the face of population growth and increasing demand for energy, particularly electricity. Microgrids have an increasing role in meeting the needs for electricity in such scenarios.

Using this model, the future increases in carbon emissions are avoided by baselining carbon emissions and implementing policies, strategies, and technologies that would level greenhouse gas emissions in the short term and reduce them in the long term. Solutions include switching to fuels that emit less greenhouse gas (e.g., coal to natural gas), implementing additional energy efficiency measures, improving energy conservation efforts, and using renewable energy. Improvement alternatives in the solution toolbox have been referred to as *wedges* as each implemented solution has an impact on GHG reduction within the stabilization triangle [19]. When implemented, these solutions would ultimately reduce the impacts of climate change (Figure 2.5). Key strategies are summarized below:

Reduce use of targeted high carbon fuels: decrease or stop the use of the targeted fuel; implement energy conservation programs.
Improve efficiencies: improve energy management; implement energy efficiency programs and efficiency improvement projects; improve the efficiency of vehicles.

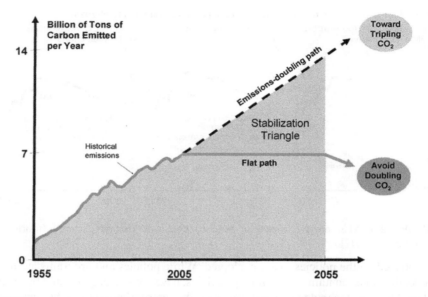

FIGURE 2.4 Graphic of stabilization triangle as applied to atmospheric carbon emissions. (Source: E&E, Princeton University [19].)

Reduce emissions from point sources: such as from power plants, stationary sources, mobility sources.

Eliminate emissions: find ways to eliminate dependence on carbon pollution generators.

Price in externalities: consider the external costs of fuel selection in project costs. This necessitates quantifying the cost of the impacts of pollution over time.

Substitute fossil energy sources: creatively switch from high carbon to lower carbon fossil fuels (e.g., from coal to natural gas).

Substitute fossil energy sources for renewables: switch to energy sources that are carbon-neutral (e.g., from coal or natural gas to solar, wind, geothermal).

Capture carbon and apply for process use: use pre-combustion or post-combustion capture to remove, collect, and store CO_2 and apply it to economically beneficial uses (e.g., enhanced oil recovery, beverage manufacturing, or carbon fiber production).

Implement carbon sequestration: provide geologic, pre-combustion, post-combustion, or biological capture and store the CO_2 in a safe location.

OTHER ENVIRONMENTAL CONCERNS

There are additional environmental concerns that are important considerations. The Clean Air Act, last amended in 1990, requires the EPA to set National Ambient Air Quality Standards (40 CFR part 50) for pollutants considered harmful to public health and the environment [21]. The Clean Air Act identifies two types of national

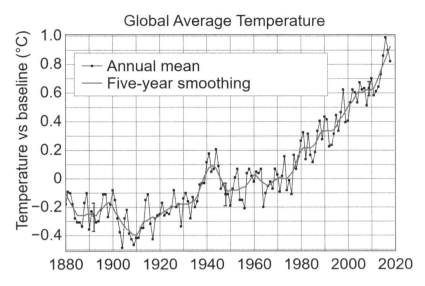

FIGURE 2.5 Changes in atmospheric temperatures, 1880 to present. (Source: NASA [20].)

ambient air quality standards. *Primary standards* that provide public health protection which include protecting the health of sensitive populations such as asthmatics, children, and the elderly [21]. *Secondary standards* provide public welfare protection, including protection against decreased visibility and damage to animals, crops, vegetation, and buildings [21]. Criteria pollutants include carbon monoxide (CO), lead (Pb), nitrogen dioxide (NO_2), ozone (O_2), particulate pollution, and sulfur dioxide (SO_2) [21]. These pollutants can harm the health of individuals, negatively impact the environment, and cause property damage. Of the six pollutants, particulate matter and ground-level ozone are the most widespread health threats. Most of these criteria pollutants are associated with fossil fuel combustion processes (see Table 2.2).

Categories of environmental contaminates include the following:

Air pollution: smoke, nitrogen oxides, sulfur oxides, ozone, carbon monoxide, carbon dioxide, volatile organic compounds, lead, particulates, and acid rain. This includes non-combusted natural gas emissions and other trace emissions from combustion processes.

Water pollution: sulfuric acid in lakes and streams, mercury, toxic metals including cadmium, arsenic, nickel, chromium, and beryllium. This includes runoff from agricultural chemicals. Effects include events such as fish kills, alga blooms, and environmental damage due to oil spills.

Land laid to waste: strip mines (over one million acres awaiting reclamation), deforestation, ash storage, lead concentrations in soils, loss of wildlife habitat. Also includes land abandoned due to nuclear accidents and storage of waste materials.

TABLE 2.2
Air Pollutants, Primary Sources, Health and Environmental Concerns (Adapted from Air Quality Lab [22])

Pollutant	Primary sources	Human health concerns	Environmental Concerns
Ozone (O_3)	Motor vehicles, exhausts, industrial processes, gassoline vapors and chemicals.	Respiratory problems, lung damage and respiratory illnesses.	Disripts photsynthesis, damages plants and reduced crop yields.
Nitrogen Oxides (NO_x)	Motor vehicles, exhausts, industrial processes, fossil fuel combustion and electric utilities.	Heart attacks, bone marrow production, asthema and nausea.	Smog, acid rain, global warming, air pollution, water pollution, toxic chemical production.
Sulfur Oxides (SO_x)	Mining operations, exhausts, industrial processes, fossil fuel combustion transportation and electric utilities.	Repiratory problems.	Acid rain, water pollution, crop damage, damage tree foilage, growth, and decreased visibility.
Carbon Monoxide (CO)	Motor vehcles exhausts, industrial processes, wood combustion, fires, and construction	Damage to cardiovascur system, damage to central nervous system and death.	Low-level smog.
Particulate matter	Motor vehicles, factories, stone crushing, wood	Respiratory issues and death	Visibility, air and water pollution.

MICROGRIDS ARE ONE SOLUTION

What are the solutions? To meet the growing need for energy by reducing the use of carbon-based fuels (while simultaneously reducing CO_2 emissions), the challenge must be met using the other tools in our engineering toolkit: energy efficiency, energy conservation, fuel substitution, renewable energy. Microgrids can synergize these tools to aid in the transition to a low-carbon economy.

Microgrids have a key role in helping meet sustainability goals. They can be configured to use local energy resources and focus on workable forms of renewable generation. With their local focus, they can be successfully deployed to meet local policy objectives. The technologies used in microgrid configurations can be designed based on locally available skills and resources. If fossil fuels are used, new microgrids are unlikely to use coal to generate electricity. Due to lower costs and dispatch capabilities, they are more likely to use natural gas to resolve intermittency problems. They have the ability to reduce greenhouse gases by enhancing our ability to use renewable energy and energy storage technologies to generate electricity. By displacing fossil fuels, they can support a GHG stabilization triangle while reducing air and water pollution.

CONCLUSIONS

Carbon emissions generated by mankind's activities are contributing to potentially irreversible changes in the world's climate. The scientific consensus is that global climate change is one result. The sustainability of life on Earth is at stake. Natural processes have effectively stored large quantities of carbon for eons and there are currently no scalable substitutes for these natural processes. The carbon compounds that are emitted into the atmosphere come from natural and human-made sources. The primary anthropogenic sources are due to the combustion of fossil fuels. The Hydrocarbon Age is at an impasse. We must find ways to use carbon-based fuels more efficiently and effectively and find workable substitutes.

We know that energy, environment, and economy are linked, yet a consensus regarding how to effectively deal with this relationship remains elusive. There is no single, quick-fix solution since greenhouse gas emissions vary widely across countries and economic sectors. Long-term solutions that reduce GHG emissions are needed, yet they are costly and difficult to implement. Policies that focus on reducing carbon emissions must be broadly based. The Kyoto Treaty and the Paris Agreement provide examples but have yielded mixed results. Programs must be implemented to achieve the goals established by those policies that we choose to adopt. Technologies that are directed toward the reduction of carbon emissions must be adaptable and economically viable. Microgrids provide a potentially viable solution as they can readily incorporate renewable energy and energy storage systems.

In the future, CO_2 is likely to become a regulated pollutant in the U.S. There are ways to improve our buildings and transportation systems that can lead to reduced GHG emissions. The initiatives in the U.S. are having a positive impact. After three years of declines, U.S. carbon emissions from energy consumption in 2017 were the lowest since 1992, the year that the United Nations Framework Convention on Climate Change came into existence [23].

There are a number of greenhouse gases more potent than CO_2 that offer emission reduction opportunities. Among the most profitable approaches is the destruction of nitrous oxide, which has 310 times the global warming potential of CO_2 and which stems from fertilizer and nylon production. Next is avoiding the release of methane gas, at least 21 times as potent as CO_2, from landfills, coal mines, and gas-flaring projects.

Ultimately, the greatest weapon in the battle against human-induced global warming will be technologies to cut CO_2 emissions from energy generation. The sooner clean energy approaches, such as renewables and near-zero fossil fuel technologies, can be refined and replicated, the more likely we can meet the challenge of reducing GHG emissions by 80% by 2050. The U.S. is making improvements toward such a goal. Due to increases in wind and solar capacities, renewable energy resources are gaining an increasing share of generation [17]. While nuclear generation was relatively flat over the past decade, it remains one of the largest sources of electrical generation without CO_2 emissions [17]. Renewable energy sources combined with nuclear energy provided about 33% of total U.S. electricity production in 2015 [17].

Renewable energy resources can be used to provide distributed generation and be widely deployed by developing microgrids. Microgrids provide the potential to resolve the principle problem of delivering energy to remote areas. They provide opportunities to decentralize and democratize the process of supplying electricity. Millions of people who lack access to electricity for basic services can develop microgrids for local use at costs that are often much less than providing grid access. The opportunity to develop microgrids provides more expedient delivery of electrical services, often at a lower cost than developing central utility systems.

We find ourselves at an impasse and solutions will be both technologically challenging and costly to implement. Solutions include fuel substitution, improving energy efficiency of vehicles, upgrading buildings and facilities, reducing emissions caused by electrical energy production, using more efficient appliances, and expanding the use of alternative energy. The potential can be measured in how policies, programs, and technologies are effectively synchronized and implemented to reduce fossil fuel consumption and carbon emissions. Microgrids, by improving efficiencies, relying more on renewables, and reducing transmission costs are seen as part of the solution to the emissions problems caused by fossil fuels.

REFERENCES

1. Jiankun, H. (2018, November 26). A holistic response – China to promote global collaboration at UN climate change event in Poland. *Time*.
2. Friedman, T. (2008). *Hot, Flat and Crowded*. Farrar, Straus and Giroux, page 214.
3. Energy Information Administration (2007, May). International energy Outlook 2007. http://www.eia.doe.gov/oiaf/ieo/emisions.html, accessed 3 November 2007.
4. Ricter, F. (2019, December 2). The global disparity in carbon footprints. https://www.statista.com/chart/16292/per-capita-co2-emissions-of-the-largest-economies, accessed 27 December 2019.
5. Adams, E. Sustainable energy, blog comments. https://insteading.com/blog/fossil-fuel-use-carbon-dioxide-emissions, accessed 2 December 2018.
6. Wikipedia. *Methane*. The chemical formula for this process is $CH_4 + 2O_2 \rightarrow CO_2 + 2H_2O$. http://en.wikipedia.org/wiki/Methane, accessed 8 March 2018.
7. Watts, A. (2019, February 19). China's greenhouse gas emissions rising at 'alarming rate'. https://wattsupwiththat.com/2019/02/05/chinas-greenhouse-gas-emissions-rising-at-alarming-rate, accessed 24 January 2020.
8. Rapier, R. (2019, December 7). The world's 10 biggest polluters. https://finance.yahoo.com/news/world-10-biggest-polluters-180000965.html, accessed 8 December 2019.
9. The World Bank. CO_2 emissions (metric tons per capita. https://data.worldbank.org/indicator/EN.ATM.CO2E.PC, accessed 17 May 2020.
10. McKinsey and Company (2007, December). Reducing U.S. greenhouse gas emissions: How much at what cost? U.S. greenhouse gas abatement mapping initiative, executive report. http://www.mckinsey.com/clientservice/ccsi/greenhousegas.asp, accessed 3 February 2008.
11. FOXNews.com (2009, April 17). EPA takes first step toward regulating pollution linked to climate change. http://www.foxnews.com/politics/first100days/2009/04/17/epa-takes-steps-regulating-pollution-linked-climate-change/, accessed 29 April 2009.
12. Lumbley, M. (2018, November). Beyond politics, a project manager's argument for clean energy. *North American Clean Energy*, 12(6), page 64.

13. SourceWatch. AES shady point generation plant. https://www.sourcewatch.org/index.php/AES_Shady_Point_Generation_Plant, accessed 24 January 2020.
14. Shin, L. (2009, February 18). EPA review reverberates through U.S. energy industry. http://solveclimate.com/blog/20090218/epa-review-reverberates-through-u-s-energy-industry, accessed 29 April 2009.
15. The Electricity Forum. (2008). Kansas lawmakers plan to deal with coal plants again. https://www.electricityforum.com/news-archive/dec08/Kansastoreturntocoalplantquestion, accessed 24 January 2020.
16. Sontakke, M. (2016, January 16). Natural gas-fired power plants are cheaper to build. https://finance.yahoo.com/news/natural-gas-fired-power-plants-210502117.html, accessed 8 February 2020.
17. Wirman, C. (2016, May 13). Carbon dioxide emissions from electricity generation in 2015 were the lowest since 1993. U.S. Energy Information Administration. https://www.eia.gov/todayinenergy/detail.php?id=26232, accessed 2 December 2018.
18. Brinkerhoff, N. and Biederman, D. (2015, January 9). Coal companies gain federal subsidies by selling coal to themselves. http://www.allgov.com/news/where-is-the-money-going/coal-companies-gain-federal-subsidies-by-selling-coal-to-themselves-150109?news=855323, accessed 9 February 2020.
19. E&E (2007, January 30). Princeton profs drive 'wedges' into policy debate. Climate Repair, a Technology Toolbox. https://www.eenews.net/special_reports/climate_repair, accessed 24 November 2019.
20. NASA (2018, January 19). GISS Surface Temperature Analysis (v4). http://data.giss.nasa.gov/gistemp/graphs, accessed 12 December 2019.
21. U.S. Environmental Protection Agency (2016, December 20). AAQS table, criteria air pollutants. https://www.epa.gov/criteria-air-pollutants/naaqs-table, accessed 2 December 2018.
22. Arteta, J. Air Quality Lab. AP Environmental Sciences. https://apeslabsjl.weebly.com/air-quality-lab.html, accessed 2 December 2018.
23. Watts, A. (2018, June 16). Chart of the week: the U.S. is the leader in CO_2 emissions reduction. https://wattsupwiththat.com/2018/07/16/chart-of-the-week-the-us-is-a-leader-in-co2-reduction, accessed 24 January 2020.

3 The Roots of Microgrids

THE EARLY HISTORY OF MICROGRIDS

Small electrical grids were the norm when electricity was first harnessed. The first local electric grid was the Manhattan Pearl Street Station in New York City, located at 255–257 Pearl Street in Manhattan [1]. Built by Thomas Edison in 1882, it began with 85 customers and powered 400 electric lamps [1]. It also pioneered two other solutions that are linked to contemporary microgrid development. It was the world's first combined heat and power (CHP) plant [2]. The steam used to power the generators was used by local manufacturers to heat nearby buildings and it also had batteries to store electricity [2]. Edison's system provided the first model of incremental microgrid growth. By 1884, the Pearl Street Station had added three more generators and expanded to serve over 500 customers with 10,164 lamps [1]. Its small but centralized generation system defined a model that eventually grew into what we know as a traditional electric grid.

The Edison Illuminating Company developed similar direct current (DC) grids in Shamokin, Pennsylvania (1882), Sunbury, Pennsylvania (1883), Brockton, Massachusetts (1883), Mount Carmel, Pennsylvania (1883), and Tamaqua, Pennsylvania (1885) [1]. As demand for electricity grew, the primary problem from the consumer perspective was obtaining access. If a facility was not in close proximity to the source of generation, it was difficult to get electricity due to the costs of transmission. By 1886, Edison's company had successfully installed 58 independent electric generating systems that we would today call microgrids. The technologies we use today for electric grid interconnections were unimportant at the time as there were no electric grids to connect to.

Hydropower Microgrids

The world's first hydroelectric power plant began operation in September 1882 on the Fox River in Appleton, Wisconsin [3]. The plant, later named the Appleton Edison Light Company, was initiated by Appleton paper manufacturer H. J. Rogers, who had been inspired by Thomas Edison's plans for an electricity-producing station in New York [2]. Unlike Edison's New York plant, which used steam power to drive its generators, the Appleton plant used the natural energy of the Fox River to produce about 12.5 KW. The operation's water wheel, generators, and copper wiring took only a few months to install and test [3]. When the plant opened, it produced enough electricity to light Rogers's home, the plant itself, and a nearby building. Such configurations were similar to what we now call microgrids.

The growing demand for electricity and the relative ease of generating hydroelectricity caused the process to be rapidly cloned. By 1886, there were 45 hydroelectric power plants in the U.S. and Canada; by 1889, there were 200 in the U.S. alone [4]. By 1920, about 40% of the electricity in the U.S. was produced by hydropower plants.

The World's Largest Microgrid in 1883

There was a series of fairs held in Louisville, Kentucky from 1883 to 1887. Called the Southern Exposition, it was a large development held in what is today called the Old Louisville neighborhood and south of today's central business district (see Figure 3.1). The exposition, held for 100 days each year on 45 acres (180,000 m^2) immediately south of Central Park (now the St. James-Belgravia Historic District), was essentially an industrial and mercantile show [5]. Though Louisville had a population of just 156,000, in the first 88 days, nearly a million people attended the exposition [6]. The exposition was larger than any previous American exhibition with the exception of the Centennial Exposition held in Philadelphia in 1876 [5]. The exhibition building was so large that it contained a half-scale Mayan temple [6].

The Southern Exhibition's electrical system included all of the features that we would find in a contemporary microgrid configuration operating as a standalone distributed generation system. It contained a generation source (see Figure 3.2), loads, transmission, etc. A highlight of the exhibition was the world's largest installation of incandescent lamps that were used to illuminate the exposition at night [5]. The lamps were invented by Thomas Edison, a previous Louisville resident. Edison's electrical system, which he personally supervised, initially included 4,600 incandescent lamps (16 candlepower each) for the exhibition hall and 400 for an art gallery; the total was more than all the electric lamps installed in New York City at that time [5]. The exterior was illuminated using Jenny arc lights. An electric trolley system (nicknamed "the electric railroad") encircled the structure. Within a few years, Louisville had one of the best electric trolley systems in the country. George H.

FIGURE 3.1 Rendering of the Southern Exposition showing electric rail line [6].

The Roots of Microgrids

Yater writes in his book *200 Years at the Fall of the Ohio*: "The Exposition was the first large space lighted by incandescence and many electrical pioneers felt that the Louisville success did more to stimulate the growth of interior electric lighting than any other Edison plant" [5].

ALTERNATING CURRENT

It soon became apparent that a large centralized power plant was more efficient at providing electricity over a wide area (and less expensive to construct) than small power plants were at providing electricity over a regional area [1]. Higher system efficiencies could be achieved using hydropower rather than fossil fuels. By the turn of the century, Westinghouse built a hydroelectric power plant on Niagara Falls, and using alternating current (AC) technology sent the electricity 20 miles to Buffalo, New York [1].

Alternating current had a distinct advantage over direct current when transportation of electricity over long distances was needed. It was much easier and cheaper to *step up* and *step down* voltage [1]. This is due to Ohm's law: current = voltage ÷ resistance, or I = V/R. When voltage increases, the resistance in the transmission line is reduced. As resistance increases, more electricity is lost as heat. The higher the voltage, the smaller the wire that can be used. As a result, DC electric generators had to be located within a mile of the load, while AC generators could be constructed much further away [1]. Most of today's modern microgrids are AC systems.

COMBINED HEAT AND POWER

A co-generation facility produces electric energy and forms of thermal energy (such as heat or steam) that are used for industrial, commercial, heating, or cooling purposes, through the sequential use of energy [8]. *Co-generation* is broadly defined as

FIGURE 3.2 Southern Exposition electric generator [7].

the coincident or simultaneous generation of usable heat and power in a single process [8]. The co-generated power is typically in the form of mechanical or electrical energy [8]. The power may be totally used in the industrial plant that serves as the host of the co-generation system, or it may be partially or totally exported to a utility grid [8]. Most CHP systems use steam turbines, internal combustion engines, and packaged co-generation systems as prime movers. Natural gas turbines are widely used in industrial plants.

MILL TOWNS AS A MODEL FOR TODAY'S MICROGRIDS

Co-generation was a model for the factory towns or mill villages of the U.S. and England in the early 1900s. These settlements clustered near mills or factories, such as cotton mills, steel works, or textile factories. The ability to offer heat for machinery enabled production. The advent of electrical lighting systems extended working hours, especially during the shorter winter days. Since water was needed in large quantities to produce steam, mill towns such as those in New England huddled near waterways (e.g., the Housatonic, Merrimack, Nashua, and Saco rivers). Many were called *mill towns* or *company towns*. An example is Pullman, Illinois, which was conceived as a place where workers could both live and work. This town was designed on the premise of being a model industrial town, complete with parks and a library [9]. All power sources were *co-located* in places where the factory's electrical loads could match the ability of the co-generation systems to produce power. These were models for today's microgrids.

Modeled on company towns, distributed generation (DG) has emerged as the next generation of CHP through relatively small distributed units (from a few kilowatts to 25 MW) [8]. DG plants may be interconnected to a utility grid or operate as standalone systems (see Figure 3.3). Typically, they are smaller, self-contained, power generation systems located near or within the boundaries of a CHP user or consumer facility and use natural gas, liquefied petroleum gas (LPG), kerosene, or diesel

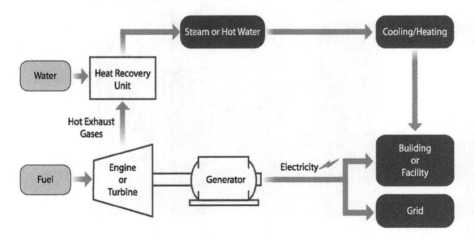

FIGURE 3.3 Combined heat and power generation schematic [11].

fuel [8]. Small-scale or micro CHP systems are available with units that provide heat to power engines which in turn drives an electric generator [10]. The recovered heat can also be used to generate additional electricity in a different process. These are called *co-generation combined cycle plants.*

CHP generates electricity and captures the heat that would otherwise be wasted to provide useful thermal energy, such as steam or hot water, that can be used for space heating, cooling, domestic hot water, and industrial processes [11]. This improves overall system efficiency. CHP co-generation systems have operated successfully in times of crisis, enabling facilities to operate independently. Modern systems can be combined with renewable generation, improving capacity. During the power outages that resulted from Hurricane Sandy, Princeton University's CHP (15 MW) and solar (5.3 MW) microgrid maintained power on campus for three days in October 2012 [12].

SHIPS ARE MICROGRIDS

Nearly all modern cruise ships and ocean liners are powered by electricity. Marine vessels are excellent models for microgrids due to their extensive use of power and electronically interfaced loads and sources [13]. Marine microgrids are central electrical networks that include generation, storage, and critical loads, and are able to operate connected to a grid (when in port) or operate in islanding mode (when out to sea) [14]. Characteristics of marine microgrids are similar to islanded land-based microgrids, except they have large highly dynamic loads, such as propulsion loads [13]. The presence of such loads and sources with power–electronic converter interfaces creates the need to manage power quality [13]. Electricity powers the motors that turns the propellers. It also powers the heating and air conditioning systems, lights, and appliances used aboard the ship. Most ships produce their electricity with turbines using diesel, natural gas, a combination of diesel and gas, or electric engines (see Figure 3.4).

Multiple microgrid technologies are being used for maritime microgrids. An example is droop control to properly share and balance the power injected by the generators [14]. The state of the art is to employ large AC/DC hybrid maritime microgrid systems, which allow integrating AC generators, AC and DC energy storage systems, and AC and DC critical and noncritical loads [14]. However, some maritime microgrids are trending more toward DC distribution systems (a DC microgrid), especially when incorporating energy storage systems [14]. Typical configurations for ships include diesel generation, an energy storage system, electric propulsion, thrusters, an internal transmission system, a power electric substation, converters, electric loads, a central control station, and a communication network. Some have small solar PV arrays to provide supplemental DC power. Excluding the propulsion system and thrusters, these are similar conceptually to the primary components found in most land-based microgrids.

Nuclear submarines are powered by nuclear reactors. Inside these reactors, a neutron is deployed to split uranium atoms. The atomic structure of uranium, once split, forms a huge amount of heat and gamma radiation. This heat emitted is used to heat water, generate steam, and drive the turbines. The turbine directly drives the

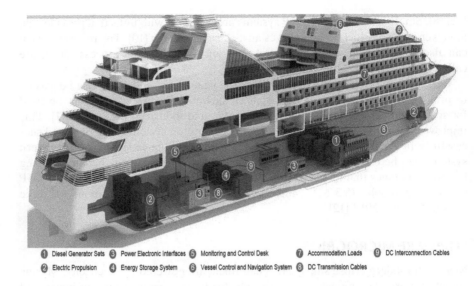

FIGURE 3.4 Cruise ship microgrid system using an on-board energy storage system [15]. (Source: https://www.et.aau.dk/digitalAssets/141/141055_cruise_vessel.bmp.)

propulsion system and electric generators; this is why they are often called "electric boats." The process is similar to the reactor systems used in commercial power plants but at a smaller scale. It is theoretically possible that the reactor from an Ohio class nuclear submarine (which has two turbines rated at 45MW each) could provide enough electricity for a village or small town.

BUILDING-SCALE NANOGRIDS

Buildings that generate their own electricity serve as small-scale models for microgrids. Initially designed to satisfy the need for reliable sources of power in the event of grid failures, hospitals and civil emergency support centers were provided with secondary power sources, often in the form of diesel generators. Many used legacy switch technologies for grid-connection to enable generators to connect and disconnect from the grid [16].

Today, many residences have backup power supply systems that might use generators or battery storage systems. Green buildings today have renewable electrical generation systems, especially solar PV, that provide electricity generated on-site.

SUSTAINABLE BUILDINGS

Buildings have an important role in reducing carbon emissions since they account for large amounts of energy use. They require interrelated systems that are energy intensive in their construction and operation. Buildings use energy for lighting, heating spaces and water, mechanical uses, appliances, and other purposes.

The Roots of Microgrids

Buildings today are being constructed with features that consume fewer resources and enable operation with less energy use. Sustainable buildings (often called green buildings) are notable due to their higher performance and enhanced capabilities as compared to standard construction practice. They are a response to past resource-intensive construction practices used in building designs with comparatively greater energy and water consumption, adverse environmental impacts, and higher costs of ownership. Sustainable buildings address issues such as site development, recycling, energy use, water consumption, renewable energy, and indoor air quality. Their primary goals include providing higher quality construction with healthy and comfortable indoor environments. They are designed to have lower operating and maintenance costs. The use of energy is a key design consideration for the construction and development of green buildings. Recently, sustainable buildings have focused on carbon emission reductions.

There are many opportunities in architectural practice to include green design features and components in buildings that address their sustainability. Developing a green building project is a balancing process that often requires tradeoffs. It involves considering how buildings are designed and constructed at each stage of the project development and delivery process. The designs typically augment the use of insulation, carefully detail fenestration, and control air and moisture infiltration. Green buildings tend to make greater use of renewable energy (see Figure 3.5).

The standards for green construction are constantly evolving. The International Energy Conservation Code (IECC) requires that certain energy efficient design methodologies be used in construction practices. It addresses the design of energy-efficient building envelopes and installation of energy-efficient systems while emphasizing system performance. It is comprehensive and provides regional guidelines with specific requirements based on climate zones. New construction materials and products are available that offer improved design solutions. Many local governments have developed their own codes that incorporate sustainable design requirements.

FIGURE 3.5 Rooftop solar collectors for residences in Amersfoort, Holland.

GREEN BUILDING ASSESSMENT METHODS

Numerous assessment systems for sustainable buildings are in use throughout the world though BREEAM and LEED set the dominate standards. The Building Research Establishment Environmental Assessment Method (BREEAM), one of the world's longest established methods of assessing, rating, and certifying the sustainability of buildings, has been used in over 70 countries. The Leadership in Energy and Environmental Design (LEED) program, originated in the U.S., considers a wide range of sustainable construction features, components, and project types. As of the end of 2019, there were approximately 70,000 LEED projects registered in the U.S. They have been completed in over 150 countries. LEED projects credit design attributes including energy and water consumption, environmental criteria, and use of renewables and green building standards.

When required by regulation, IECC requirements are included in BREEM or LEED projects. This holds the building designers using these certification programs to higher energy-efficiency standards. Many green building technologies such as high-efficiency windows, solar arrays, and day-lighting applications are readily observable. Regardless, it is sometimes discouraging to owners that many important engineered features of green buildings are hidden from view in mechanical rooms and inaccessible spaces. Examples of these technologies include enhanced insulation, lighting control systems, computer control systems to manage energy and water use, rainwater collection systems, under-floor airflow systems, and geothermal heating and air conditioning systems, among others. LEED certification programs encourage the use of such generation technologies by providing credits for identified percentages of renewable generation and purchased renewable power.

Energy managers and engineers recognize the importance of sustainable buildings, their primary components, and the importance of resource and energy use in green construction practice. They are often the leaders in promoting green design features in the lighting, control, electrical, and mechanical systems that are found in a growing number of buildings. Green buildings advance the use of renewable energy for on-site electricity generation, as many have solar PV arrays and small-scale wind power systems.

Validating green building performance is challenging due to the subjective concepts involved, the evolving nature of standards, and the local variability of construction practices and regulations. Measurement and verification of building performance and resource consumption has an important role in this process. It provides procedures that verify the energy use, water consumption, and emissions reductions green buildings offer. As improved performance is verified, the use of green construction techniques and technologies will further expand as more examples of successful green buildings become available.

ZERO NET ENERGY BUILDINGS

Many building owners recognize the importance of sustainable buildings and of resource and energy use in green construction. They are among the leaders in promoting green design features in the lighting, electrical, and mechanical systems that are found in a growing number of buildings.

Many green building designers attack the problem of energy usage in two ways. First, the buildings are designed to be more energy efficient as compared to buildings designed to local standards. Second, they design the buildings to take advantage of renewable energy either by on-site generation or through purchase of green power. On-site generation typically includes use of renewables, such as solar energy or geo-exchange systems, that further reduce the need to purchase energy from local utilities. Buildings that power themselves have revolutionized how buildings function yet may disrupt the electric utility industry in the future as the practice becomes more widespread. Building owners that utilize green power purchase programs essentially partner with the utility that provides renewable electricity.

Some newer structures are designed to be zero net energy (ZNE), net positive energy (NPE), or carbon-neutral buildings. ZNE buildings represent a new category of green buildings and a departure from the perspective that a building can be considered green without generating on-site power. ZNE buildings have been completed in Japan, Malaysia, the U.S. and elsewhere. U.S. examples of ZNE buildings include the NREL Research and Support Facility in Golden, Colorado, and the North Shore Community College Health and Services building in Danvers, Massachusetts.

It would be better if buildings could generate more power than they require and routinely sell excess power back to the grid. Perhaps some can. Such buildings are called net positive energy buildings. In the UK, a model energy-positive project, called The Active Office, produced more than 150% of the energy it consumed [17]. This project was funded by Innovate UK, with support from Swansea University and the European Regional Development Fund. By combining a range of innovative technologies, it generates, stores, and releases solar energy in one integrated system [17]. Technologies include a roofing system with integrated solar cells, a solar photovoltaic thermal system capable of generating both heat and electricity, lithium ion batteries for electricity storage, plus a 2,000-liter (528-gallon) water tank to contain solar heat [17]. Energy-positive buildings lower energy costs for the consumer and eliminate the need for peak central power generating capacity [17].

Nanogrids

Building-scale ZNE and NPE microgrids are often called *nanogrids*. They are indifferent to whether or not electricity from the host utility is available and functioning (see Figure 3.6). Typically serving a single building or a single load, nanogrids are very small microgrids [18]. Navigant Research defines a *nanogrid* as being 100 kW for grid-tied systems and 5 kW for remote systems not interconnected with a utility grid [18]. Since nanogrids are restricted to a small and limited service area, they are not constrained by regulations prohibiting the transfer or sharing of power across a public right-of-way [18]. They have a greater propensity to use direct current systems.

The components of a residential nanogrid might include renewable energy systems, energy storage, a building energy management system (BEMS), and a smart electric meter. The BEMS monitors, automates, and controls building systems such as heating, ventilation, air conditioning, thermostats, and lighting to increase

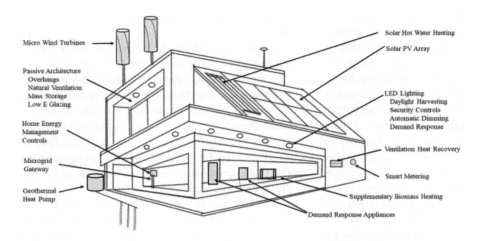

FIGURE 3.6 Energy-efficient residence designed as a nanogrid. (Concept adapted from [20].)

building energy efficiency [19]. They can monitor electrical energy generated by the renewable energy systems (e.g., solar PV and wind power) and in conjunction with smart metering systems make decisions as to how to manage energy use in the residence. The primary benefits are reduced energy costs, on-site electricity generation, improved comfort and interior environmental conditions, and the ability to remotely program and monitor conditions. Residential nanogrids may also incorporate charging stations for automobiles or other vehicles.

Many residences—almost 26 million households or an estimated 100 million people—are served by off-grid renewable energy systems: some 20 million households through solar home systems, 5 million households by renewables-based minigrids, and 800,000 households by small wind turbines [21].

The U.S. Virgin Islands and Puerto Rico were hit by hurricanes Irma and Maria during 2017. These were severe storms which disrupted power for months. Some residents decided to rebuild their homes and incorporate new solar PV systems in an effort to improve reliability (see Figure 3.7). Those with larger systems have essentially constructed nanogrids. The question is whether or not these systems will survive the high wind conditions associated with hurricanes and continue to generate electricity afterward.

REGIONAL GRIDS

There is another category of electric grids for those too large to be considered microgrids yet not large enough to be considered to be similar to central grids. These systems are characterized by multiple load centers, and low or medium voltages for transmission, yet are not connected to a central national grid. Holdmann and Asmus have proposed the definition for *regional microgrid* as "a high voltage transmission network connecting multiple distribution nodes/load centers and power stations, but that is entirely isolated from a larger national or continental central grid or is only weakly connected" [22]. An attribute of regional microgrids is that they align with

The Roots of Microgrids

FIGURE 3.7 Residence using solar PV on St. Thomas Island, U.S. Virgin Islands.

distributed energy resources management systems (DERMS) [22]. These types of control systems manage the centers of electrical energy production to match the loads of the regional systems.

The majority of the people living in Alaska, Canada, and the Russian far east have electric power supplied by regional grids that connect generation sources and load centers via a transmission network [22]. Most Arctic regional grids are anchored by renewable generation systems, such as geothermal and hydropower, along with a base load primary energy source such as diesel or natural gas generation [22]. Examples of regional grids that rely heavily on hydropower include the Railbelt and Southeast Alaska Power Agency grids in Alaska, and the Yukon, Taltson, and Snare Grids in Canada [22].

The Nordic country of Iceland offers an example of a regional microgrid. Being a remote island disconnected from the Nordic Grid, it is often considered to be too large of a system to be considered a microgrid. The Landsnet Transmission Grid in 2017 was powered by 14,059 GWh (73.3%) of hydropower, 5,170 GWh (26.6%) of geothermal energy, and only 2 GWh (.01%) of fossil fuel-fired generation [22].

Norway's electricity grid is configured with three levels: the nationwide transmission grid (operated by Statnett), the regional grid, and the distribution grid [23]. The transmission grid makes it possible to transmit electricity from the hydropower plants in the southwest and north to consumers in other parts of Norway and to nearby countries. Regional grid companies, engaged in production and electricity trading, own about 6% of Norway's transmission grid. They are owned by private companies, municipalities, and county governments. Both the distribution and regional grids are considered distribution systems, as defined by EU legislation [23]. Its regional grids typically link the transmission grid to the distribution grid, and may also include production and consumption radials that carry higher voltages [23]. The regional grid carries voltages from 33 kV to 132 kV, and has a total transmission length of about 19,000 km [23]. The distribution grid consists of the local electricity grids that normally supply power to smaller end-users [23]. It carries a voltage of up to 22 kV, divided into high-voltage and low-voltage segments with the low-voltage distribution to ordinary customers normally carrying 230V or 400V [23].

THE ROLE OF TRANSPORTATION SYSTEMS

In 2018, the U.S. transportation industry consumed the equivalent of 28.3 quadrillion Btus or 27.9% of its total of 101.3 quadrillion Btus of primary energy [24]. Plans to reduce greenhouse gases (GHGs) must include ways to reduce impacts from the combustion of transportation fuels. Most of our mobility fuels are derived from fossil fuels, primarily petroleum (92%), and to a lesser extent natural gas (3%) and renewables (5%) [24]. Products include diesel fuels, gasoline, and ethanol. Combustion of petroleum fuels creates environmental problems by adding GHGs, pollutants, ozone, and soot to the atmosphere. While mobility fuel use continues to increase, we have other options. Solutions that decrease these pollutants include reducing the use of vehicles by planning solutions, manufacturing vehicles that are more efficient, using alternative means of transportation, and using fuels that are less polluting. Fuel substitution is another. These can lead us to a more sustainable energy future when petroleum fuels are replaced with renewables.

Planning solutions include infrastructure improvements and densification. Cities can be designed to minimize the impact of automobiles, discourage vehicle use, provide green wedges from the city center to its outskirts, and by encouraging brownfield redevelopment. Well-designed urban planning approaches encourages people to walk, bicycle, and skateboard, by locating people in close proximity to urban amenities and creating transit centers. Infrastructure improvements such as bikeways, bus rapid transit, light rail, and waterways have a place in sustainable planning.

We can provide more efficient vehicles. This has been a focus of successful legislative efforts to improve fleet vehicle mileage requirements. It has been accomplished by improving the efficiency of the combustion engines used in automobiles and trucks and by reducing vehicle weights and sizes. Vehicular electrification technologies provide opportunities for microgrids to be developed to support personal transportation systems.

Alternative means of transportation enable consumer choice by providing multiple transportation options. Arriving at the central train station in Amsterdam by bicycle, canal boat, bus, tram, or taxi, you can connect to rail service that will get you anywhere in Europe. While vehicles are not allowed to park at the station, there is a spacious three-story garage for bicycles.

The railways in The Netherlands operate on electricity generated primarily by wind power. Yet renewable fuels in the U.S. comprise only a small portion of its transportation fuels. Examples include biogas-derived compressed natural gas, ethanol, hydrogen vehicles using fuel cells, electric vehicles that are charged using renewable energy, and many others. In Atlanta, one of the largest landfills produces methane gas which is used as a fuel for garbage trucks. The main commercially

The Roots of Microgrids

produced biofuel in the U.S. is ethanol, created using corn. In most processes, the corn kernels are extracted for this purpose as their sugar content is higher than in the other bulky materials, such as the husks and stalks. Producing cellulosic ethanol from these waste materials is more costly but provides a higher energy production ratio than from the kernels and is similar to that of using sugar cane. A biopower plant that generates 2.5 MW for a locality of 4,000 households has been proposed using sorghum to ferment ethanol for combined heat and power [25]. This novel combined ethanol and power microgrid model has more robust and stable characteristics when compared to conventional microgrids [25].

Providing alternatives to fossil fuels for transportation systems is challenging. Planning solutions include infrastructure improvements and densification but applications are limited once infrastructure is in place. Alternative fuels must complete with gasoline and diesel fuels which have remarkably high energy production ratios. However, GHG emissions from the combustion of petroleum fuels are substantial. Not so with biofuels which recycle the carbon dioxide they extract from the atmosphere. Many researchers believe algae many be a future solution. When grown under specific conditions, they produce oils plus byproducts that can be used to make ethanol. Compared to cultivating corn, growing algae requires less land area and uses less water, pesticides, and energy. Carbon emissions from the process are actually negative.

Solar PV-powered vehicle charging stations are becoming more common in the U.S. Some are designed for fleet mobility uses. A solar array was constructed over an existing parking lot at the Cincinnati Zoo and Botanical Gardens primarily to produce electricity for the facilities. It is configured as a microgrid and also provides electricity for electric vehicle charging stations (see Figure 3.8). Another example is the City of Asheville, North Carolina, which has electric-powered police cars and

FIGURE 3.8 Solar array at the Cincinnati Zoo and Botanical Gardens.

its own vehicle charging stations. Today, there are many other examples of solar PV being used to supply electricity to the batteries found in electric vehicles. Using fuel cell vehicles is another alternative. Las Vegas, Nevada has its own hydrogen fueling station for hydrogen-powered vehicles.

We need synergistic, multi-faceted approaches to resolving our energy-related transportation emissions. It begins with rethinking the design of our transportation infrastructure, offering more efficient vehicles, more options, and greater use of renewable energy systems.

SUMMARY

The generation of electrical power began with hydropower and diesel generation systems. It is reasonable to claim that Thomas Edison was one of the early developers of local microgrids. Microgrids have been called a back-to-the-future solution and are considered to be smaller versions of the electric grid [25]. Microgrids are a proven way to provide power to remote areas and locations that lack access to established electrical supply networks [26]. These first microgrids were developed in the U.S. in the early 1880s. The world's largest microgrid in 1883 provided power for lighting and electric trollies for an exposition center in Louisville, Kentucky. It was the first example of a microgrid serving both building and transportation systems. Early in the history of electrical generation, both AC and DC electrical generation systems were available. If a DC electric load was not close to the source of generation, it was difficult to supply electricity due to the costs of transmission. Ultimately, AC systems became the preferred alternative since they could more efficiently transmit electricity over larger geographic areas. It is interesting that this problem persists to modern times, and remains a driver for microgrid development.

Models for microgrids include ships and building-scale nanogrids. Both have internal energy provision systems that provide all or a portion of their electricity requirements. Marine examples include combined heat and power systems for on-board systems using diesel generation when out to sea. Green building programs are used to certify a building's sustainable features including the use of renewable energy systems. Some buildings are designed to harness on-site sources of energy, such as solar energy, for electrical or thermal generation and can operate in stand-alone mode. Zero net energy buildings generate power on-site for building use. Many are interconnected with the electric grid, providing excess electricity when available. Transportation systems have a role in supporting the development of microgrids. Electrification of transportation systems provides opportunities to support microgrid development since vehicular batteries can be charged using renewables and serve as storage devices.

Regional electric grids are usually too large to be considered microgrids yet not large enough to be considered similar to central grids. These systems are characterized by multiple load centers and low or medium voltages for transmission, yet are not connected to a central national grid.

Microgrids have a varied and notable history which provides many successful real-world examples of successful applications. It is apparent that what we today call a traditional electric grid grew from the notion of microgrids that were designed to provide electricity at a local scale. Yet microgrids present a different set of challenges than traditional grids. Microgrids are smaller and highly customized, and require specialized designs. For microgrid solutions to be successful, engineering expertise is required to design, develop, and maintain microgrids.

REFERENCES

1. JS (2012, October 25). How electricity grew up? A brief history of the electrical grid. https://power2switch.com/blog/how-electricity-grew-up-a-brief-history-of-the-electrical-grid, accessed 1 November 2018.
2. LO3 Team (2018, February 16). All about microgrids. https://lo3energy.com/find-energy-industry-needs-microgrids-work-now-work-future, accessed 24 January.
3. World History Project. The world's first hydroelectric power plant begins operation. https://worldhistoryproject.org/1882/9/30/the-worlds-first-hydroelectric-power-plant-begins-operation, accessed 1 November 2018.
4. Hull, R. (2015, October 6). Electric light by water. https://news.asce.org/electric-light-by-water, accessed 1 November 2018.
5. Wikipedia. Southern exposition. https://en.wikipedia.org/wiki/Southern_Exposition.
6. Historic Photos of Louisville, Kentucky. The southern exposition 1883–1887. Originally published in *Harper's Weekly*. https://historiclouisville.weebly.com/the-southern-exposition-1883-1887.html, accessed 1 December 2019.
7. Historic Photos of Louisville and Environs. https://historiclouisville.weebly.com/the-southern-exposition-1883-1887.html.
8. Wong, J. and Kovacik, J. (2018). Cogeneration and distributed generation, in *Energy Management Handbook*, 9th edition (Roosa, S., ed.), The Fairmont Press, pages 171–191.
9. VCU Libraries. Company towns: 1880s to 1935. https://socialwelfare.library.vcu.edu/programs/housing/company-towns-1890s-to-1935, accessed 11 June 2019.
10. Lekule, S. Lekule blog. Electricity and technologies. http://sosteneslekule.blogspot.com/2015/09/hybrid-power-generation-systems.html, accessed 10 November 2018.
11. U.S. Environmental Protection Agency. What is CHP? https://www.epa.gov/chp/what-chp, accessed 4 December 2019.
12. North Carolina Clean Energy: Technology Center. *Serving Critical Infrastructure with Microgrids, 16 Combined Heat and Power and Solar PV*. North Carolina State University.
13. Gamini, S., Meegahapola, L., Fernando, N., Jin, Z. and Guerrero, J. (2017, February). Review of ship microgrids: system architectures, storage technologies and power quality aspects. http://vbn.aau.dk/en/publications/review-of-ship-microgrids-system-architectures-storage-technologies-and-power-quality-aspects(e62c6b16-5415-4813-b5ac-c1d2bdc0b043).html, accessed 24 January 2020.
14. Aalborg University. Marine microgrids. https://www.et.aau.dk/research-programmes/microgrids/mission-and-focus-areas/maritime-microgrids, accessed 24 November 2018.
15. Aalborg University. Marine microgrids. https://www.et.aau.dk/digitalAssets/141/141055_cruise_vessel.bmp.
16. U.S. Department of Energy (2011, August 30). DOE microgrid workshop report. https://www.energy.gov/oe/downloads/microgrid-workshop-report-august-2011, accessed 19 January 2019.
17. Fortuna, C. (2018, July 5). Energy-positive buildings can become power stations. *Green Building Elements*. https://greenbuildingelements.com/2018/07/05/energy-positive-buildings, accessed 17 May 2020.

18. Hardesty, L. (2014, March 25). What is a microgrid? https://www.energymanagertoday.com/whats-a-nanogrid-099702, accessed 24 January 2020.
19. Killough, D. (2014, August 4). Building energy management systems save energy and money. *Green Building Elements*. https://greenbuildingelements.com/2014/08/06/building-energy-management-systems-save-energy-money, accessed 17 May 2020.
20. Nexus Energy Center. http://www.nexusenergycenter.org/home-indoor-air-quality-guide, accessed 6 December 2019.
21. International Renewable Energy Agency (2015). Off-grid renewable energy systems: status and methodological issues, working paper. https://www.irena.org/DocumentDownloads/Publications/IRENA_Off-grid_Renewable_Systems_WP_2015.pdf, accessed 18 January 2019.
22. Holdmann, G. and Asmus, P. (2019, September). What is a microgrid today? *Distributed Energy*, page 16.
23. Norwegian Ministry of Petroleum and Energy (2019). The electricity grid. https://energifaktanorge.no/en/norsk-energiforsyning/kraftnett, accessed 24 January 2020.
24. Energy Information Administration (2019). Monthly energy review.
25. Zhang, L., Zu, X., Fu, J, Li, J. and Li, S. (2019, December). A novel combined ethanol and power model of microgrid driven by sweet sorghum stalks using ASSF. *Energy Procedia*, 103, pages 244–249.
26. Roosa, S. (2019). It's back to the future with microgrids. *International Journal of Strategic Energy and Environmental Planning*, 1(4), page 5.

4 Traditional Electrical Supply Systems

SUPPLYING THE ELECTRIC GRID

A conventional power station, also referred to as a power plant, powerhouse, generating station, or generating plant, is an industrial facility for the generation of electric power. Most power stations contain one or more generators, a rotating machine that essentially converts mechanical energy into electrical power. Conventional power stations, such as coal-fired, gas, and nuclear-powered plants, hydroelectric dams, and large-scale solar power stations, are centralized and often require electricity to be transmitted over long distances [1]. Central plants typically rely on one type of fuel for generation. Distribution systems provide power from the plant to the customer using a one-way model for power transmission and flow. These plants, owned by large utility companies, are used to supply power to the electric grid. Their chief objective is to supply uninterruptable power to their customers in the exact quantities instantaneously required.

WHAT IS THE ELECTRIC GRID?

An *electrical grid* is an interconnected network for delivering electricity from suppliers (typically a utility) to consumers. The existing electrical system we call the *grid* began as a set of connections among various standalone systems. Today we might call these standalone distributed systems microgrids. The grid consists of generating stations that produce electrical power, high-voltage transmission lines that carry power from distant sources to demand centers, and distribution lines that connect to the individual customers [2]. Generation sources for traditional grids produce electricity which is instantaneously supplied and instantaneously consumed.

To enable use of the electricity, it needs to travel from the source of generation to the point of consumption. This process is called transmission. Transmission equipment and power lines supported on transmission towers transfer the bulk electricity using high voltages. These voltages are used to electrically transfer larger amounts of energy with lower transmission line losses, since resistance in the transmission wires is often the greatest cause of system loses. The primary transmission (high voltage) lines continue to receiving stations that step down the voltage to secondary transmission (low voltage) networks. Distribution equipment components include relays, interrupters, distributors, and service mains which provide electricity directly to consumers. Users consuming the electricity are typically categorized as residential, commercial, and industrial loads. These loads might include power for lighting, motors, machinery, thermal applications, buildings, or transportation systems.

National grids supply power to entire countries and *transnational* or *international grids* are interconnected with other countries, typically geographically adjacent to one another [2]. These grids are vast. The U.S. electrical grid is comprised of 7,300 generating stations, 55,000 substations, plus 160,000 miles (257,000 km) of power lines which are owned and operated by over 500 companies [3,4]. This system provides power to 145 million discreet consumers. Due to its massive scale and millions of components, it has been called the world's largest and most complicated machine (see Figure 4.1). Much of the electrical generation relies on fossil fuels. For example, the U.S. has about 400 coal-fired power plants (down from 589 coal plants in 2011) that generate 30% of the nation's electricity [5]. In 2017, there were 27 U.S. states that relied on coal to produce at least 25% of their electricity [5].

For North America, the *grid* is actually two major and three minor alternating current (AC) power grids that are called *interconnections* [6]. The U.S. segments of these grids that serve the lower 48 states are the Eastern Interconnect (east of the Rocky Mountains to the Atlantic and part of northern Texas), the Texas Interconnect (Electric Reliability Council of Texas, ERCOT), and the Western Interconnect (area from the Rocky Mountains to the Pacific). These three interconnections operate independently and have only limited power exchanges among them [4]. The network structure of the interconnections improves system reliability by providing multiple routes for power to flow and allowing generators to supply electricity to multiple load centers [4]. Both the Eastern Interconnect and Western Interconnect actually operate as transnational grids since they accept and supply electrical power to nearby provinces in Canada.

Renewable energy is energy derived from natural, non-depletable, or rapidly replenished resources—sun, wind, water, tidal, biological processes, and geothermal heat [7]. With the exceptions of tidal (due to gravitational forces) and geothermal (from the Earth's internal heat) energy, all energy traces its origins back to the sun. Solar energy from the sun's radiation may be converted into heat or electrical power [7]. Wind energy technologies harness the kinetic energy of air movement caused by solar radiation. Biomass processes store energy from solar radiation as chemical bonds in organic matter through the process of photosynthesis, and may then be passed through the food chain or converted into thermal, kinetic, and electrical energy through a variety of conversion processes [7].

CATEGORIES OF ENERGY RESOURCES

Fossil fuels are forms of energy formed from deposits of compressed and heated organic matter that consist primarily of carbon and hydrogen bonds. Common types include peat, coal, lignite, petroleum, natural gas, and other bitumen-based fuels. Uranium is also a fossil fuel. A key feature of fossil fuels is that they are considered to be depletable energy resources. Once they are used in a combustion process they typically cannot be reused. The exception is uranium as there are ways to recondition spent fuel rods, enabling them to be recycled and reused. However, such processes are not used in North America.

Today's alternative energy resources are often distinguished by low environmental impact in contrast to fossil fuels. Sometimes considered synonymous with

Traditional Electrical Supply Systems 55

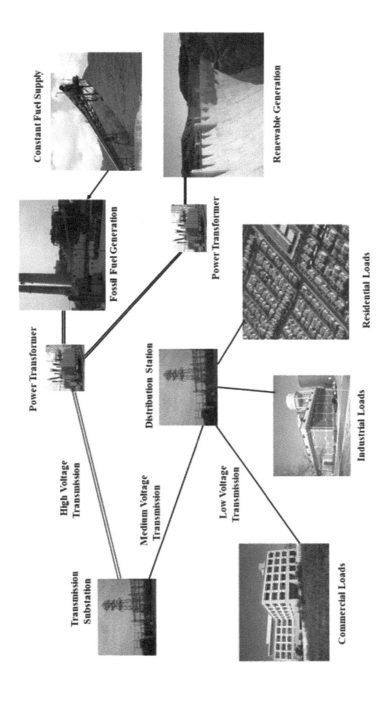

FIGURE 4.1 Basic structure of the electrical power system. (Source: adapted from [8].)

renewables, alternative energy is, in fact, a broader class of resources and energy conversion technologies that include energy efficiency, combined heat and power, and zero-emissions conversion technologies such as fuel cells [7]. Fundamentally, *alternative energy* is simply the substituting of one form of energy for another to achieve a purpose such as lowering costs or environmental emissions. Examples include switching from coal to natural gas-fired systems or using solar generated electricity rather than nuclear power.

The definition of clean energy is more ambiguous and variable. Generally, *clean energy* is considered to be any type of energy, such as renewables or nuclear power, that does not cause atmospheric pollution; this is opposed to fossil fuels, such as coal or oil, that exhaust pollution when combusted [9]. Some experts believe that nuclear power should be categorized as a clean energy source since harnessing nuclear energy does not directly emit greenhouse gas (GHG) emissions; others believe that nuclear power should not be included as a clean energy source because of the radioactive nuclear wastes that are generated when producing electricity [10]. If clean energy refers only to energy sources that do not produce GHG emissions then nuclear power should be considered a clean energy option; otherwise, if clean refers to energy sources that are not hazardous for the environment then nuclear power cannot be categorized as a clean energy source [10]. Others define clean energy as energy that is derived from inexhaustible, zero-emissions sources including renewables and energy saved through energy efficiency measures, yet excluding nuclear power [11]. To further confound the definition, a broad interpretation sometimes includes natural gas in the mix of clean energy resources, despite it being a fossil fuel, since it can be combusted more efficiently, and releases less nitrogen and CO_2 into the atmosphere, than coal and oil.

TRADITIONAL SOURCES OF GRID-SUPPLIED ELECRICAL POWER

The traditional ways of generating grid-supplied electricity are discussed in this section. Excluding hydropower, most of these approaches use thermal plants whose processes involve combustion of fossil fuels to generate electricity. Traditional electrical grids are characterized by a one-way flow of electricity from the center of production to the consumers. They also tend to be dependent on traditional resources to generate power, specifically fossil fuels. Along with initiatives toward energy efficiency and associated conventional technologies, renewable energy resources must be deployed to effectively reduce GHGs in a sustainable manner.

LARGE HYDROPOWER

Hydropower provides an important way to generate electricity with a renewable resource. It was the primary means of generating electricity using renewables in the U.S. until the late 20th century. However, these projects are hydrology-dependent and a set of natural and geophysical conditions must be available at a selected site for it to be economically feasible. Of the 80,000 existing dams in the U.S., only 2,400 currently generate hydroelectricity [3]. These generate power when the pressure of falling water spins turbines, rotating a shaft that causes generators to produce electricity.

Traditional Electrical Supply Systems 57

In the U.S., utilities have proposed 70 new projects to upgrade existing generating facilities or to build new capacity. These projects when completed are anticipated to increase hydropower electrical production by 11,000 MW over the next decade [3].

Large hydropower systems are developed as major public works projects. The methodology for generating electricity is dependable, scalable, and highly efficient. Most plants use a dam to hold back water and create a reservoir. The intake gates on the dam are opened allowing gravity to cause water to flow though the penstocks (pipelines leading to a turbine). The penstock might be tapered over its length, wider at the intake and narrower at the exit point near to where the flowing water strikes the blades of the turbine. Gravity increases the water pressure as it flows though the penstock. The water strikes the blades of the turbine, which rotates and is attached by a turbine generator shaft to the generator located above it. The most common turbine used in hydropower generation is the Francis Turbine which is a large disc-shaped component with curved blades. The water pressure is so great that the turbine might weigh hundreds of tons yet rotate at a rate of 90 revolutions per minute. As the turbine generator shaft rotates, so does a series of magnets inside the generator (rotor), which rotate past copper coils (stator) and cause alternating current (AC) to be produced. The transformer inside the powerhouse increases the power's voltage, creating higher-voltage current. Exiting the power plant are four primary wires, three supplying three phases of power simultaneously and the fourth being a neutral or ground which is common to the other three. The water then flows away from the turbine through pipelines called *tailraces* which allow the water to reenter the waterway downstream.

The world's largest power station of any type is the Three Gorges Dam, located on the Yangtze River in China, which became fully operational in 2012. The project produces hydroelectricity, increases the river's navigation capacity, and reduces the potential for floods downstream by providing flood storage space. The dam's annual electric generating capacity is roughly 22,500 MW. Its construction submerged 244 square miles (632 km^2) of land and displaced over 1.5 million people. Other large hydroelectric plants include the Itaipu owned by Brazil and Paraguay (14,000 MW), and the Guri power plant on the Caroni River in Venezuela (10,300 MW).

The U.S. has 1,444 hydroelectric dams that generate 7% of the nation's grid-supplied electricity [5]. The Grand Coulee plant on the Columbia River in the U.S. state of Washington, rated at 10,080 MW, was originally constructed to generate 7,600 MW and is an example of how older plants can be upgraded to increase output. Another uprating project from 1986 to 1993 for the Hoover Dam (see Figure 4.2) in Arizona increased its capacity from 1,345 MW to 2,080 MW [12]. In 2015, the site generated 3.6 TWh [12].

COAL-FIRED ELECTRICAL POWER

Second only to natural gas, coal-fired electrical generating plants are one of the primary ways power is produced in the U.S. In 2017, the U.S. had roughly 285 GW of thermal coal-fired generation, down from 357 GW in 2008. The share of electricity generated from coal dropped from 51% in 2008, to 33% in 2015, then to 31% in 2016, due primarily to market forces and low-cost natural gas [13]. The U.S. has

FIGURE 4.2 Low water levels of the reservoir at the Hoover Dam in 2010 reduced plant electrical output.

400 coal-fired power plants, with coal producing the majority of the electrical power in 13 states [5]. The impact of the Clean Power Plan and market forces have shuttered 111 plants since 2015 [5]. Of the remaining coal plants, roughly a quarter will be retired or converted to natural gas and another 17% are considered to be uneconomical to operate [13]. This shift away from the use of coal in the U.S. has resulted in reduced greenhouse, sulfur oxide, and nitrogen oxide emissions plus substantial benefits in the form of improved health [13].

Like most fossil fuel generation, coal plants often use a conventional Rankin Cycle process in which the fuel is combusted, water is heated in a boiler where it changes into steam, and the steam pressure is used to drive a turbine generation system. The steam is then condensed and changes again into a liquid state and the water is reused. A reliable water source is required and due to the large amounts of water needed, these plants are typically located near a river or large lake. There are two types of plants: 1) once-through plants pump the water directly from the water source, heat it, and then discharge it back to the original source; and 2) wet-recirculating plants that avoid discharging the heated water by cooling and reusing the water.

Natural Gas-Fired Electrical Power

The U.S. has 1,793 natural gas power plants that generate 34% of the nation's electricity [5]. Natural gas is the primary power source in 19 U.S. states and its use in electrical generation is increasing [5]. Compared to other fossil fuel-fired electrical plants, NG plants are relatively inexpensive to construct and more easily permitted, and have higher thermodynamic efficiencies and lower associated GHG emissions. Some are designed to reuse waste heat. However, the inability to store large amounts of natural gas on-site is a limitation of natural gas generation. Some U.S. utilities purchase natural gas in the summer when demand and costs are lower, transfer it to storage facilities, and draw from storage in the winter when demand is greater and costs are higher.

There are two primary types of natural gas-fired power plants. Simple cycle plants consist of a natural gas-fired turbine connected to a generator. The simple cycle plants are less complex but less efficient than combined cycle plants. However, simple cycle plants are able to dispatch faster than coal or nuclear plants [14]. Combined cycle plants include the components of a simple cycle plant plus have an external combustion engine enabling plant efficiencies up to 60% [14]. Their combined cycle extends beyond this, using more of the energy created during combustion. The exhaust gases can be directed to flow toward the heat recovery steam generator, in which the hot gases boil pre-heated water into steam that next expand through a turbine, generating additional electricity [14]. Some plants are capable of using both oil and natural gas to generate electricity (Figure 4.3).

Diesel Generation

Reciprocating engine generators deliver reliable energy and are fully dispatchable while renewable sources are offline or producing at less than their rated capacities [16]. Modular units rated from 350 kW to 15 MW and larger can be added incrementally to create microgrids with greater capacities. The modularity of multiple units adds flexibility for variable load conditions [16]. Generator sets offer high power densities, simple-cycle electrical efficiencies from 40% to 48%, high part-load efficiency, and excellent capability to follow loads. They tolerate a broad range of ambient temperatures and operate at higher altitudes without derating. Diesel units accept loads rapidly, with start times to full load as fast as ten seconds and ramp-up times from 25% to 100% load in as quickly as five seconds [16]. They are capable of unlimited starts and stops with limited impact on service life or maintenance requirements [16]. The technology is thoroughly proven and reliable, with hundreds of gigawatts of capacity installed worldwide [16]. For these reasons, diesel generation has been a preferred technology for island and remote microgrids when storage is available and fuel transportation costs are low.

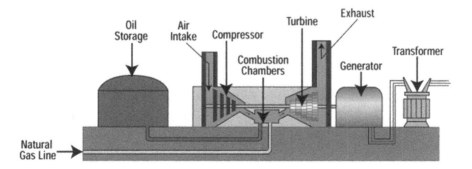

FIGURE 4.3 Design of an electrical generation system using oil and natural gas [15].

Gensets using diesel also offer the potential for combined heat and power (CHP). Indeed, applications that make microgrids economically attractive—communities, resorts, industrial facilities—tend to have significant thermal requirements [16]. Heat captured from engine exhaust or cooling circuits can be converted to steam, hot water, or chilled water (by way of absorption chillers), or used in water desalination plants. In each case, the project economics can be greatly enhanced [16]. In temperate climate zones, it is very attractive to use both CHP and solar energy because of their complementary nature [16]. CHP capabilities are typically used during cooler times of the year when the solar contribution lessens [16].

Diesel generation presents a set of infrastructure and supply-chain requirements to generate electricity, especially for remote locations. Oil production and refineries may not be locally available or nearby. In most cases, diesel fuel must be transported to the site and stored in close proximity to be immediately available for the combustion equipment. If diesel fuel is imported, tank storage is needed near port facilities in locations where land costs are likely at a premium (see Figure 4.4). For the generation process, fresh water must be available in ample quantities for steam conversion. Maintenance support for the generation site must be readily available. These logistics and support concerns raise the fuel costs, increase safety and financial risks, and must be addressed.

Nuclear Power

Nuclear power is often classified as a clean energy source as its greenhouse gas emissions are negligible. The fuel is widely available but impacted by the economics of production. Support for nuclear power is mixed. While countries such as Finland,

FIGURE 4.4 Typical island liquid fuel storage facilities.

Canada, China, and Russia are planning to increase the use of nuclear power, others including Spain, Germany, Switzerland, Belgium, and France are phasing out or reducing their shares of nuclear power production. Nuclear fuels are mined using underground mining, open pits, in-situ solution mining, or heap leaching [17]. Preparing the fuel for use involves several complicated front-end processes, including milling, conversion, enrichment, reconversion, and nuclear fuel fabrication [17]. For many countries, it is less expensive to import fuel rods than to manufacture them domestically.

The U.S. has more nuclear power plants than any other country in the world, followed by France. As of 2018, there were 61 nuclear power plants (36 having two or more reactors) with a total of 99 operating nuclear reactors in 30 U.S. states [18]. Of the 31 countries in the world with commercial nuclear power plants, the U.S. has the most nuclear electricity generation capacity and has generated more electricity from nuclear energy than any other country [18]. France has the second highest nuclear electricity generation capacity and obtains 78% of its total electricity generation from nuclear energy, the largest share of any country (see Table 4.1) [18]. There are 14 other countries that generate at least 20% of their electricity from nuclear power [18]. Almost all nuclear-generated electricity is used to supply central grids.

Nuclear power plants create thermal energy by nuclear fission, boiling water to produce steam. The fission reaction takes place inside the nuclear power plant's reactor core which contains uranium fuel formed into energy-rich ceramic pellets [18]. These pellets are stacked to form metal fuel rods 12 feet (3.7 m) long [18]. A bundle of the fuel rods, some with hundreds of rods, is called a *fuel assembly* [18]. Using a Rankin Cycle process the heated water is converted to high-pressure steam which expands and is used to rotate large turbine generators to generate electricity. Nuclear plants cool the steam back into water in a cooling tower located at the power plant or use cooler water pumped from nearby ponds, rivers, or the ocean. Baseload nuclear plants typically have capacity factors that exceed 80% [19].

In the U.S., after the spent fuel rods are used they must be stored on-site as the U.S. has no permanent storage location to handle high-level nuclear wastes. Some states have specific regulations which prohibit the transfer of nuclear wastes through

TABLE 4.1
Capacity of Nuclear Power Stations

Country	Nuclear electricity generation capacity (million kW)	Nuclear electricity generation (billion kWh)	Share of country's total electricity generation
U.S.	99	797	19.5%
France	63	419	77.6%
Russia	25	183	18.1%
China	27	161	2.9%
South Korea	22	157	30.4%

Source: USEIA, International Energy Statistics, August 8, 2018.

their jurisdictions. Billions of dollars of federal funds have been expended to prepare a site for underground storage of nuclear waste at Yucca Mountain in Nye County, Nevada. However, after years of development the site has not yet been licensed to store nuclear wastes. Transportation of spent fuel rods across jurisdictions is not permitted in most U.S. states. Currently, power stations in the U.S. must store nuclear wastes in dry steel and concrete casks on-site and may be required to do so for thousands of years. As an alternative, the European Union has a facility to recondition spent fuel rods for reuse. However, the number of times that the fuel rods can be recycled are limited.

PROBLEMS WITH THE ELECTRIC GRID

Electric grids throughout the world are aging. The oldest systems are inefficient and utility and transmission companies face high costs in their efforts to maintain them. Electric grid reliability varies widely throughout the world (see Figure 4.5). Though it has served the country well, the U.S. electric grid includes segments of the system that are over 100 years old. Roughly 70% of the U.S. grid's transmission lines and power transformers are over 25 years old, and the average age of its power plants is over 30 years old [6]. Large electric grids are difficult to secure, expensive to operate, and arduous to maintain [20]. In some cases, replacing segments with more modern modular technologies is less expensive from a lifecycle cost perspective than attempting to maintain the existing equipment and infrastructure [21]. Electric demands on the grid have increased along with rising domestic consumption.

The strategy used to prevent and manage service interruptions has been to design the systems to include redundant infrastructure if possible. When outages occur in segments of transmission lines, they are bypassed until they can be repaired. Yet this strategy has proved costly. To prevent future power outages much of the transmission system and its supporting equipment need to be substantially upgraded.

SECURITY ISSUES

In the past, electric utility companies have focused on the physical security of their management centers, generation equipment, substations, and transmission infrastructure. This focus is no longer adequate. Security threats to the utility grid include actors such as nation-states, politically motivated hackers, disgruntled utility employees, and many others [21]. Successful attacks on segments of the grid can cause financial losses plus impact the safety of company employees and the public. Power outages caused by cyberattacks have local and national security implications. Deliberate power outages, such as those in California in 2019 that were intended to prevent fires due to dry conditions and high winds, can unintentionally ignite wildfires when transmission lines are reenergized causing billions of dollars in damage.

As electrical generation systems become increasingly managed by internet-connected direct digital control systems, vulnerabilities tend to intensify [21]. The potential for cyberattacks is increasing and the ones that occur are becoming increasingly sophisticated. "Since at least March 2016," reads an alert from the United States

Traditional Electrical Supply Systems

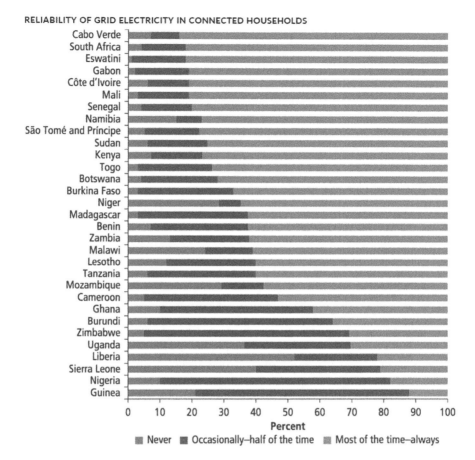

FIGURE 4.5 Electrical grid reliability in developing countries [28].

Computer Emergency Readiness Team (US-CERT), "Russian government cyber actors [...] targeted government entities and multiple U.S. critical infrastructure sectors, including the energy, nuclear, commercial facilities, water, aviation and critical manufacturing sectors" [22]. It further stated:

> This campaign comprises two distinct categories of victims: staging and intended targets. The initial victims are peripheral organizations such as trusted third-party suppliers with less secure networks, referred to as staging targets throughout this alert. The threat actors used the staging targets' networks as pivot points and malware repositories when targeting their final intended victims."

These initial intrusion efforts involved a scouting mission after gaining access and did not result in loss of control or outages. However, in December 2015, a Russian group used similar tactics to shut down 30 electrical substations in western Ukraine, cutting power to 230,000 citizens in the first known successful cyberattack on a central power grid [23].

Securing the electric grid involves restricting physical access to plants, safeguarding networks with firewalls and data encryption, and maintaining system integrity by keeping unauthorized actors from making changes to the system [23]. Establishing and maintaining cybersecurity safeguards is not easily accomplished since much of the utility system equipment is not state-of-the-art and each power station is unique. Each utility substation is also a potential point of attack. The objective of the safeguards is to create a situation that requires potential attackers to overcome multiple time-consuming and difficult obstacles in order to access utility company computer systems [23].

System Inefficiencies and Power Outages

The existing electric distribution system is inefficient. This is due to aging equipment, obsolete system layout, outdated engineering, and management obstacles that include the difficulties of operating a vertically integrated industry in a deregulated environment [2]. The system creates high rates of outages along with the potential for cascading failures. Many electrical power plants exhaust two-thirds of the energy in the source fuel as unusable heat. Distribution systems in the U.S. waste another 2% to 13% of the energy, as losses are magnified by power transmission [24]. Losses ranging from 8% to 10% are typical.

Inefficiencies include power interruptions, some only for seconds, with extended outages lasting for weeks or months after severe weather events. In the U.S., climate change is increasing the risk of damage to petroleum, natural gas, and electrical infrastructure along the east and Gulf coasts from rising sea levels and hurricanes of greater intensities [25]. More frequent and intense precipitation events are anticipated that increase the risk of floods for coastal and inland energy infrastructure [25]. Higher ambient temperatures lower the operating thermal efficiencies and generating capacities of thermoelectric power plants, reducing their efficiencies and the current-carrying capabilities of transmission and distribution lines [25]. Due to climate change, the U.S. energy supply system is projected to be increasingly threatened with infrastructure damage, fuel availability, and supply imbalances that will cause more frequent and longer-lasting power outages [25].

Extreme weather can impact all components of electricity generation systems. Examples include floods, wind damage, ice, lack of rain, mudslides, etc., especially those caused by clusters of tornados and hurricanes which usually cause widespread damage to infrastructure. Electrical outages create substantial economic losses in communities impacted by extreme weather conditions. After severe weather events occur, utility companies often spend large amounts of money upgrading their infrastructure. These costs are eventually reimbursed by the ratepayers. The U.S. Department of Energy estimates that power outages cost U.S. businesses roughly $150 billion per year [26]. Power outages create unreliable electric grid connections, which are a concern in North America but more so in Africa and parts of Asia where it is difficult to mobilize response efforts.

Other drawbacks include costs to install and maintain step-up and step-down transformers, obtaining rights-of-way, legal expenses, and the cost of settlements due to outages [26]. Generating capacities require that large amounts of excess capacity be

Traditional Electrical Supply Systems

built into the electrical grid, increasing system investment and maintenance costs [27]. Grid management needs constant monitoring and management, requiring balanced electrical generation with electrical demand satisfied over large geographic areas [27]. Despite such problems grid reliability tends to be taken for granted in the developed world. Reliability is much less ensured in developing countries.

ENVIRONMENTAL ISSUES

A fundamental feature of the electric grid is that it is comprised mostly of large power plants that use fossil fuels for generation (hydropower plants are the exception). Utility companies with fossil fuel-fired generation expend large sums of money, supported by taxpayers and ratepayers, to mitigate the environmental consequences of the air and water pollution they create. Much of these expenditures, often to control sulfur and nitrogen oxides and perhaps mercury, have not been amortized. Some have problems with ash storage ponds and concerns about long-term containment. The Kingston Fossil Plant in Tennessee experienced a failure of an ash pond dam in 2008 that released over a billion gallons of slurry, flooding surrounding land and causing hundreds of millions of dollars in damage. The cleanup costs paid by the Tennessee Valley Authority (TVA) ultimately totaled more than a billion dollars and the project required eight years to complete.

Utility companies find themselves in a quandary as to how to increase electrical production without increasing carbon emissions [29]. As the use of funds from public utilities becomes increasingly transparent, consumers observe utility expenditures, funded by rate payers, being used to impede the institution of the environmental regulations they are advocating. The widely advertised mantra of *clean coal* as a solution has proved too costly to implement, is not scalable, and yet is widely advocated by utility companies using ratepayer funds [29]. Costs for such plants are so prohibitive and undefinable that in some parts of the U.S. even utility support is waning. Yet the fundamental environmental issues with fossil fuel-fired electricity generation remain mostly unchecked, and continue to cause human health problems in populations living near fossil fuel plants.

To add to the conundrum, the fossil fuel plants use large quantities of fresh water, which is becoming increasingly costly. To extract fresh water at low cost, the plants are typically located near lakes and rivers where the resource is considered to be free. Often the actual value of extracting and pumping water are hidden costs not found on the expense sheets of many utilities. Valuing water to improve sustainability is a robust and detailed process that requires measurement, valuation, decision-making, and governance [30]. If this or a similar process is not followed, the value of water is externalized. Watersheds throughout the world are experiencing unsustainable amounts of water extraction that degrade water quality and fossil fuel plants are part of the problem [30].

REGULATION OF THE ELECTRIC GRID

Electric distribution systems have required regulation since the 1930s. Due to periodic regulatory changes, electric utilities have a history of focusing on compliance.

Regulation happens at the generation, transmission, and distribution levels of the system (see Figure 4.6). Utility companies routinely attempt to influence political processes that enable them to maintain a monopoly in their regulated territories of operation. Using a purely socialist business model with oversight by public service commissions, most are able to recover all their costs and are guaranteed a profit. They are protected from default even when gross mismanagement occurs and laws are violated. When fines are paid due to regulatory control, rate structures are often adjusted to cover the increased costs.

The electrical power system in the continental U.S. consists of three mostly distinct large interconnections: the Eastern Interconnection, the Western Interconnection, and the ERCOT. Each operates as an integrated machine that directs the flow of power and instantaneously matches generation with loads. Regulation of the national grids is balkanized and complicated [31]. Each state can regulate many aspects of the electric power system within their jurisdictional boundaries [31]. Typically, the regulations are codified by the state legislature and carried out by a public utility commission which oversees retail rates, distribution reliability, and the services of investor-owned electric utilities [31]. The Federal Energy Regulatory Commission (FERC) regulates the transmission and sale of electricity in interstate commerce.

CONNECTING MICROGRIDS WITH THE GRID

Connecting a microgrid to an existing electric power system (EPS) requires a legal contract between the microgrid owners and the EPS owners [33]. The technical design must ensure the safe, reliable, and economic operation of both the microgrid and the macrogrid [33]. Among other requirements, the point of interconnection (POI) between the microgrid and the EPS may require automatic islanding, synchronization, and dispatch controls [33]. Both active and reactive power dispatch is

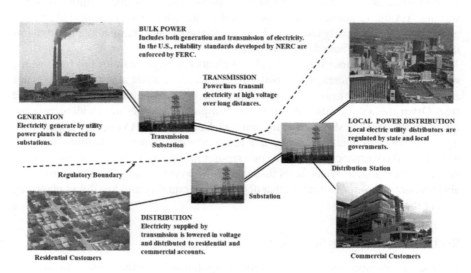

FIGURE 4.6 Electric grid distribution and regulation model. (Source: adapted from [32].)

typically performed by the relay logic of the POI, which may simultaneously dispatch multiple DERs [33]. This makes a microgrid that is configured with multiple DERs appear to be a single dispatchable generator to a utility [33]. A series of field tests of the interconnection and its control technologies is often needed to insure safe and reliable operation.

Advanced microgrid controllers facilitate microgrid connections to the main grid. The improved capabilities of the controllers have led to greater interest in the possible applications of grid-tied microgrids [34]. When a microgrid is grid-connected, its distributed generators and electrical storage system (if any) will synchronize to the frequency and magnitude of the grid voltage, optimizing the energy supply as determined by the energy management unit [34]. In island mode, intelligent microgrid controllers manage and maintain the steady state and dynamic power balance between load, generation, and energy storage without dependence on the grid or communications infrastructure [34]. Utilities have struggled to create business models that enable microgrid deployment at larger scales [35]. Recent studies are targeting how microgrids can be standardized for wider adoption and provide more benefits to the larger utility grid [35].

SUMMARY

The strategy of traditional electrical supply systems has been to provide power to consumers by linking centralized power plants to the grid with high-voltage transmission combined with low-voltage distribution systems [36]. The often-used analogy is that the power stations provide electricity to fill the *lake* and consumers are the *straws* that pull electricity from the lake. Traditional power generation sources include hydropower, coal-fired generation, diesel generation, natural gas-fired generation, and nuclear power, mostly dependent on fossil fuels. The number of coal-fired power plants is declining in developed countries such as the U.S. and Germany but increasing in many developing countries such as China and India. In the U.S., a key driver in reducing coal consumption has been greater shale gas extraction from fracking which has created large supplies of comparatively inexpensive natural gas [37]. Over the past decade, consumption of renewable electricity in the U.S. rose by 349 terawatt hours (TWh), yet power from natural gas increased by 696 TWh—almost twice the renewable energy contribution [37]. Supplies are so large that the U.S. no longer requires substantive imports of natural gas from Canada. The availability of low-cost natural gas in the U.S. has been a boon for microgrids that depend on the fuel to smooth generation from intermittent renewable energy sources.

Service interruptions that create large-scale power outages are becoming more frequent and larger in scope. A power outage on Manhattan Island in New York City that occurred in June 2019, putting 73,000 residents in the dark, was caused by a single transformer fire [38]. In August 2019, a National Grid failure impacted the southeastern United Kingdom and much of London. Trains, vehicular traffic, and residences were impacted by the outage. Securing the electric grid involves restricting physical access to plants and safeguarding networks with cybersecurity capabilities. The costs required to maintain central grids increases their costs, creating markets for microgrid development.

For certain consumers and businesses, regardless of where they are located, a reliable electric supply is critically important [39]. Some businesses have equipment that must operate continuously or risk substantial losses and restart costs. Microgrids are seen as a potential solution, preventing unscheduled shutdowns. Many traditional sources of electrical generation provide models for microgrids yet the scale of electrical production for microgrid is typically much smaller. As utilities incrementally update their transmission and distribution systems, regulators implement new approaches to unlock innovative microgrid business models, and large customers (e.g., municipalities and industrial facilities) become more economically and logistically driven to ensure their energy security, microgrids will become increasingly attractive [38].

Microgrids are a solution as they are capable of being connected to their host grids and providing more resilient electrical power. Microgrids that are grid-connected synchronize their distributed generators and storage to the frequency and magnitude of the grid voltage to optimize the energy supply. Microgrids, when properly designed, can ensure high electrical system reliability while also reducing fuel costs and emissions [39]. They can provide viable local alternatives to grid power if needed.

REFERENCES

1. Wikipedia (2018, October 17). Distributed generation. https://en.wikipedia.org/wiki/Distributed_generation, accessed 2 November 2018.
2. Wikipedia (2018, October 17). Electric grid. https://en.wikipedia.org/wiki/Electrical_grid, accessed 2 November 2018.
3. Davidson, P. (2008, October 28). Water power gets new spark. *USA Today*, page 33.
4. Desjardins, J. (2015, August 18). Mapping every power plant in the U.S. http://www.visualcapitalist.com/mapping-every-power-plant-in-the-united-states, accessed 2 November 2018.
5. U.S. Department of Energy (2014, November 17). Understanding the grid. https://www.energy.gov/articles/infographic-understanding-grid, accessed 7 November 2018.
6. Energy Information Administration (2016, July 20). U.S. electric system is made up of interconnections and balancing authorities. https://www.eia.gov/todayinenergy/detail.php?id=27152, accessed 7 November 2018.
7. Roosa, S., ed. (2018). *Energy Management Handbook*, 9th edition. Fairmont Press: Lilburn, Georgia, page 453.
8. Electrical and Electronics Engineering. Structure of electrical power system. https://esegateelectrical2u.blogspot.com/p/structure-of-electrical-system.html, accessed 7 December 2019.
9. Explore Dictionary (2018). Clean energy. https://www.dictionary.com/browse/clean-energy, accessed 25 November 2018.
10. Haluzan, N. (2010, November 22). Clean energy definition. http://www.renewables-info.com/energy_definitions/clean_energy_definition.html?jjj=1543154626206, accessed 24 January 2020.
11. North Carolina Sustainable Energy Association (2018). What is clean energy? https://energync.org/what-is-clean-energy, accessed 24 January 2020.
12. Wikipedia (2018). Hoover Dam. https://en.wikipedia.org/wiki/Hoover_Dam, accessed 17 May 2020
13. Union of Concerned Scientists (2017). A dwindling role for coal. https://www.ucsusa.org/clean-energy/coal-and-other-fossil-fuels/coal-transition#.W-SlaPZFzuk, accessed 8 November 2018.

Traditional Electrical Supply Systems

14. Afework, B., Hanania, J., Jenden, J., Stenhouse, K. and Donev, J. (2018). Energy education, natural gas power plant. https://energyeducation.ca/encyclopedia/Natural_gas_power_plant, accessed 19 November 2018.
15. Center for Environment, Commerce and Energy. cenvironment.blogspot.com, accessed 25 November 2018.
16. Saury, F. and Tomlinson, C. (2016, February). Hybrid microgrids: the time is now. http://s7d2.scene7.com/is/content/Caterpillar/C10868274, accessed 24 January 2020.
17. United States Energy Information Administration (2018, August 8). Nuclear explained, the nuclear fuel cycle. https://www.eia.gov/energyexplained/index.php?page=nuclear_fuel_cycle, accessed 20 December 2018.
18. United States Energy Information Administration (2018, August 8). Nuclear explained, nuclear power plants. https://www.eia.gov/energyexplained/index.php?page=nuclear_power_plants, accessed 20 December 2018.
19. Purdue University, State Energy Forecasting Group (2018, October). 2018 Indiana renewable energy resources study, page 16. https://www.purdue.edu/discoverypark/sufg/docs/publications/2018_RenewablesReport.pdf, accessed 17 May 2020.
20. Lumbley, M. (2018, November). Beyond politics, a project manager's argument for clean energy. *North American Clean Energy*, 12(6), page 64.
21. Didier, D. (2008, November). Space invaders – protecting America's power information systems from outside interference. *North American Clean Energy*, 12(6), page 62.
22. Cyberwarzone (2018, July 31). Russian activity against critical infrastructure. https://cyberwarzone.com/russian-activity-against-critical-infrastructure, accessed 17 May 2020.
23. Bloomberg (2018, May 15). Protecting critical infrastructure in a digital world. https://www.bloomberg.com/news/sponsors/siemens/protecting-critical-infrastructure-in-a-digital-world, accessed 2 November 2018.
24. Grimley, M. and Farrell, J. (2016, March). ILSR's energy democracy initiative. https://ilsr.org/wp-content/uploads/downloads/2016/03/Report-Mighty-Microgrids-PDF-3_3_16.pdf, accessed 26 January 2020.
25. Zamuda, C., Bilello, D., Conzelmann, G., Mecray, E., Satsangi, A., Tidwell, V. and Walker, B. (2018). Energy supply, delivery, and demand, in *Impacts, Risks, and Adaptation in the United States: Fourth National Climate Assessment, Volume II*, (Pryor, S., ed), Chapter 4. U.S. Global Change Research Program: Washington, DC, pages 178–195. https://nca2018.globalchange.gov/downloads/NCA4_Ch04_Energy_Full.pdf, accessed 17 May 2020.
26. Office of Nuclear Energy (2018, January 25). DoE report explores U.S. advanced small modular reactors to boost grid resiliency. https://www.energy.gov/ne/articles/department-energy-report-explores-us-advanced-small-modular-reactors-boost-grid, accessed 4 December 2018.
27. Zactruba, J. (2018). The advantages of microgrids. Bright Hub Engineering. https://www.brighthubengineering.com/power-generation-distribution/90436-a-system-of-systems-microgrids-poised-to-soar-in-popularity, accessed 24 January 2020.
28. Routley, N. (2019, November 27). Mapped: the 1.2 billion people without access to electricity. https://www.visualcapitalist.com/mapped-billion-people-without-access-to-electricity/?fbclid=IwAR3r3e4lwziKhreJQBFKRcY89OODJoxLz273oqFMxGYqkbaHDHAZnBS1Vjo, accessed 30 November 2019.
29. Roosa, S. and Jhaveri, A. (2009). *Carbon Reduction – Policies, Strategies and Technologies*. Fairmont Press: Lilburn, Georgia, pages 148–149.
30. Garrick, D., Hall, J., Dobson, A., Damania, R., Grafton, R., Hope, R., Hepburn, C., Bark, R., Boltz, F., Stefano, L., O'Donnell, E., Mathews, N. and Money, A. (2017, November 24). Valuing water for sustainable development. *Science*, 358(6,366), pages 1003–1005.

31. National Academies of Sciences, Engineering and Medicine (2016). *The Power of Change: Innovation and Deployment of Increasing Clean Electric Power Technologies*, The National Academies Press: Washington, DC, pages 153–161.
32. The Heritage Foundation. BG 2959, heritage.org. http://www.mondaq.com/images/profile/individual/415434a.jpg, accessed 17 May 2020
33. Manson, S. (2017, March 23). Connecting a microgrid to the grid. *Microgrid Knowledge*. https://microgridknowledge.com/connecting-microgrid-to-grid, accessed 7 December 2018.
34. Rycroft, M. (2015, October 15). The grid-connected microgrid: local network of the future? *EE Publishers*. http://www.ee.co.za/article/grid-connected-microgrid-local-network-future.html, accessed 7 December 2018.
35. Trabish, H. (2017, February 23). Utilities' microgrid strategy. *New Energy News*. http://newenergynews.blogspot.com/2017/08/original-reporting-utilities-microgrid.html, accessed 26 January 2020.
36. National Renewable Energy Laboratory (2013, April). NREL leads energy systems integration. *Continuum Magazine*, 4.
37. Rapier, R. (2019, December 7). The world's 10 biggest polluters. https://finance.yahoo.com/news/world-10-biggest-polluters-180000965.html, accessed 24 January 2020.
38. Sanchez, S. (2019, July 13). Power restored after partial New York City blackout leaves thousands without electricity. *USA Today*. https://www.usatoday.com/story/news/nation/2019/07/13/manhattan-power-outage-widespread-power-outages-new-york-city/1726521001, accessed 22 August 2019.
39. GTM Research (2016, February). Integrating high levels of renewables into microgrids: opportunities, challenges and strategies. www.sustainablepowersystems.com/wp-content/uploads/2016/03/GTM-Whitepaper-Integrating-High-Levels-of-Renewables-into-Microgrids.pdf, accessed 16 January 2019.

5 Microgrid Architecture and Regulation

FEATURES OF MICROGRID ARCHITECTURE

The critical infrastructures of today's modern world rely on having dependable electricity supplied continuously. Microgrid designers develop system architecture that attempts to achieve improved scalability, flexibility, and security. The basic architecture of a microgrid includes generation, critical and noncritical electrical loads, a central controller, power interfaces, a point of common coupling, and storage technologies. Digital systems manage interoperable distributive control functions and capabilities. The supportive architecture for microgrids has the elusive goals of ultimately providing component plug-and-play functionality, two-way electrical power transfer, and seamless connectivity with the utility grid. This chapter considers the features of microgrid architecture, their primary components, advanced microgrids, and reviews microgrid regulations and standards.

MICROGRID OPERATIONAL CONFIGURATIONS

Microgrids are configured as substructures of the *macrogrid* that are comprised of local low-voltage and medium-voltage distribution systems with distributed energy resources (DER) and storage devices that satisfy the demands of energy consumers [1]. These systems can be operated in a *semiautonomous* way, if interconnected to the grid, or in an *autonomous* way (islanding mode) if disconnected from the main grid [1]. Islanding capabilities are a fundamental feature of many microgrids. Intentional islanding is the act of physically disconnecting a set of electric circuits from a utility system, and operating those circuits independently [2]. *Anti-islanding* refers to safety protocols to prevent DERs from feeding power onto utility distribution networks during system outages [2]. These are designed to prevent harm to workers and damage to the grid during outages and power restoration efforts.

There are several basic components of the architecture of a microgrid that are fundamental. Most importantly, microgrids must contain the equipment necessary to generate electricity to satisfy loads. Microgrids usually have a form of energy storage system which does not necessarily need to be an internal system. It may satisfy this need by purchasing external storage services or allowing the host electric grid, if interconnected, to serve as the primary electricity storage system. Within the microgrid there must be a set of electricity-consuming devices that dictate the loads placed on the microgrid. Examples include lighting systems, electric motors and fans, heating and air conditioning equipment, refrigeration systems, water heating, etc.—all typically associated with facilities and buildings. Automatic control systems may be used that enable selected loads to be cycled or shed during periods of peak electrical demand to optimize load management. Finally, grid-connected microgrids require a utility interconnection to enable the microgrid to exchange power with the larger utility network. If a microgrid is designed to operate only independently in island mode, then the utility interconnection with the macrogrid is not required.

The various generation components and auxiliary equipment of microgrids can present design problems when planning microgrid architecture. Often, the multiple types of electrical generation systems are not initially designed or configured to operate in concert with other systems. Regardless of microgrid configuration, to achieve the goal of providing resilient electrical power when needed, microgrids must be designed and constructed to ensure safe and seamless interoperability among the various electrical generation sources and any connected electrical storage equipment.

There are three primary conceptual configurations for electrical utility distribution systems: centralized, decentralized, and distributed (see Figure 5.1). Centralized systems are a direct connection of nodes to a central point of electrical production. This is common in a utility configuration in which there is only one central power plant and transmission links and connections carry electricity from the plant directly to individual users (loads). The central plant provides continuous power by having multiple generators. However, a breakdown of the configuration at its central station may cause the entire system to fail. Decentralized configurations have multiple points of electrical generation which connect to their direct loads but also connect to either a central station or a number of remote stations. Decentralized configurations tend to have greater resiliency than centralized systems since they

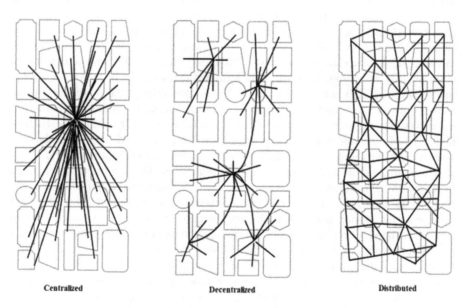

FIGURE 5.1 Centralized, decentralized, and distributed network configurations. (Source: concept adapted from Baran, P., RAND Institute [1964].)

lack a central point of failure. Distributed configurations provide both generation and loads at each node and are linked to all other generation and load points in a web-like configuration. Distributed networks are the most resilient since a breakdown that occurs at any station or link can be resolved by rerouting transmission to ensure reliability.

Microgrids typically use either decentralized or distributed network configurations within their boundaries. By using a multi-resource model as a solution to the problem of electrictrification, a diverse portfolio of production options improves system redundancy to reinforce the power grid if a single electrical generating asset or resource is compromised [3]. Distributed configurations are most resilient for situations when single transmission pathways or specific nodes are compromised.

Planning for Electrical Distribution Systems

There is an evolution in the development of electrical supply systems from central grid systems to smaller more resilient decentralized systems. From the utility's planning perspective, the aging of grid infrastructure increases the need for future upgrades and spawns rate increases. Electric utility companies, especially those with regulated service areas, may profit from the development of the improvements plus finance the costs by increasing the tariffs paid by their customers. Some electric customers, dissatisfied with the increasing costs and declining services of their utility supplier, are generating their own power and using the grid-supplied power for backup purposes [4]. From the host utility's perspective, this arrangement may not be profitable since additional capacity must be available on short notice if the customer requires large amounts of additional power. Some respond with increases in fees or fixed rates for access.

Implementing grid-related improvements is not only expensive but also creates service interruptions and major inconvenience. Often these events transfer costs to non-stakeholders who are not parties to the electrical supply transaction. In November 2018, a transmission line replacement by the electric utility LGE over a major interstate highway in Kentucky was scheduled as a two-hour event but instead required an entire day. The construction created a traffic jam over 15 miles (24.1 km) long leaving commuters stranded on remote sections of the highway for hours without access to services. Those impacted by the utility's decision were not forewarned and received no compensation for their lost time, cost of delay, or increased travel expenses.

Planning at the building scale usually focuses on the building itself without consideration of capacity optimization. When a new skyscraper is constructed in an urban area, it is often assumed that the utility will simply have additional power on hand to meet the added loads. During the design process, architects often have no idea of the building's estimated electrical usage nor have they attempted to calculate the building's electrical demand. This process often negates opportunities for load optimization and results in higher operating costs that are transferred to the owners. The building is plugged into the electric network and the hope is that the utility will have excess capacity for the long term.

Long-range electrical generation planning is capacity- and transmission capability-focused and often based on extension of regional demand projections. Building and system loads are considered to be beyond utility control. Planning at this scale often fails to account for the evolution away from central coal and nuclear plants which are being replaced by systems that generate electricity by renewable energy and natural gas sources [4]. Adjusting to intermittent resources and mixed fuel generation means that energy supply strategies must also adjust to be successful [4]. From the utility company perspective, meeting the requirements of dispersed electrical generation can be challenging.

Traditional electrical distribution systems provide a one-way transfer of electricity from the producer to its consumers. Some regional electric utility companies with regulated service territories behave as if they own the customers they serve and have sole rights to supplying power. When a microgrid is developed the question of who owns the customer becomes less clear. When customers are linked to the transmission system of the distributed energy resource (DER) developer, there can be two-way power flows between the DER and the local utility or power flow among and through DER developers. When effectively designed, system resiliency and reliability can be improved. For example, DER developers can supply power by orchestrating delivery from multiple developers through system interconnections to large numbers of consumers. Figure 5.2 shows local microgrid supply options for consumers linked to DERs.

A proposed improvement to standard grid architecture is the smart grid model. The smart grid provides a set of functional categories that supports a conceptual understanding of the grid frameworks. According to Eger et al., "From an architectural point of view the microgrid can be represented by selected viewpoints. Microgrid use cases can be decomposed into four interoperability layers modelling

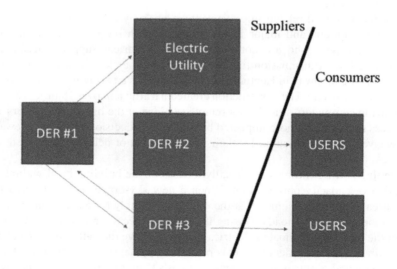

FIGURE 5.2 Local microgrid supply capability diagram for consumers using DERs as the primary electrical energy source.

use case functions, information model and data requirements, communication and connectivity requirements and the components" [5]. The architecture of microgrids consists of the following interoperability layers:

Function layer: the function layer represents use cases, functions, and services independent from their physical implementation in systems and components.
Information layer: the information layer describes the information-related requirements, objects, or data models which are required by the use cases, functions, or services.
Communication layer: the communication layer describes all the communication and connectivity requirements.
Component layer: the component layer is the physical distribution of all participating components in the microgrid context. This includes power system equipment (typically located at process and field level), protection and control devices, network infrastructure, and computer system [5].

Smart grids are structured to optimize the use of technologies that provide automated, digital, and intelligent decision-making algorithms, and use data collected from sensors monitoring these components.

ALTERNATIVE ARCHITECTURES FOR AC AND DC MICROGRIDS

For microgrids, there are several types of basic architecture, divided into either alternating current (AC) or direct current (DC) networks. Operating efficiencies of microgrid configurations will vary. Microgrid networks might include firm generation; electrical generation that is controllable and dispatchable, such as a diesel generator; combined heat and power (CHP) generation; or possibly battery storage [6]. Energy storage systems and firm generation types are dissimilar within microgrid configurations. For example, the basic architecture of AC network configurations with solar photovoltaic (PV) used as the sample generation source include the following options [6]:

AC network: the AC grid acts like a 100% efficient storage system; the timing of the generation relative to the load is not critical.
AC microgrid + PV + battery storage: when generation and load are matched and there is no charging or discharging to and from the storage system, the energy flows and efficiencies are the same as in the AC network. In the disjoint case, the storage system absorbs all energy from the PV system and then delivers that energy to the aggregate internal loads at a later time. The use of storage to manage the net flow of power leads to losses as the batteries are charged and discharged.
AC microgrid with firm generation: firm generation complicates the analysis of architecture efficiency since generator efficiency can depend strongly on loading and speed. With AC architecture, the electrical losses are from the conversion of power from the AC bus to the DC internal loads assuming the generator is operated to handle microgrid loads [6].

The basic architecture of DC network configurations with solar PV used as the sample generation source includes the following options [6]:

DC network: the bidirectional inverter/converter imposes a round trip efficiency for exchanges with the AC grid.

DC microgrid + PV + battery storage: with storage system operation schemes similar to AC microgrids, similar logic can be applied to determine the efficiencies of a DC microgrid. In the matched case, the storage system is not charged or discharged, and the operation and efficiencies are the same as the DC network during matched loads. In the disjoint case, the storage system absorbs all energy from the PV system and then delivers that energy to the aggregate DC internal loads when needed. Efficiency losses will result.

DC microgrid with firm generation: similar logic applies to the DC architecture, but the power from the generator passes first through an AC–DC conversion step to get to the DC bus, but can then be used directly by 50% of the native DC loads at high voltage [6].

COMPONENTS OF MICROGRIDS

There are primary components of microgrids which are keys to the design of their architecture. Microgrids are designed based on the need to link and manage power generation sources with connected loads to meet selected goals. Enabled by intelligent control technology, microgrids manage the operation of all linked DERs while connected to the utility grid or utilized as an independent power system [7].

Power Sources

Microgrids use distributed energy resources to generate electrical power. DERs are comprised of electrical generation and storage systems and can be deployed in a large number of units [8]. They have a number of common features: 1) they are not centrally planned; 2) they are often owned and operated by an independent power producer; 3) their power is not centrally dispatched; 4) they are interconnected to the central electric power system at any convenient point in the grid; 5) when operating connected to the grid, they may modify grid operation; and 6) the power supplied can be either dispatchable or non-dispatchable depending on configuration [8].

Microgrids can be powered by distributed fossil fuel generators, batteries, or renewable resources such as solar panels and wind turbine generators [9]. Note that batteries when discharging are seen by the system as a source of power generation.

Microgeneration is the electricity generated by homes and small business that is distributed locally [10]. Despite the granular nature of the generation, the contribution from microgeneration sources in total makes a substantial impact. In the UK, renewable energy resources account for one-third of all electricity and microgeneration accounts for 17% of the total [10]. At times renewable distributed resources are the primary source of electricity in the UK. For example, on 8 December 2019, the UK generated 16.2 GW of electricity from wind which accounted for 43.7% of its

Microgrid Architecture and Regulation 77

electricity requirement and biomass provided an additional 7.9%. While renewables accounted for 51.6% of the total electricity needed, coal supplied only 3.1%.

An advantage of microgrid design is the flexibility to select from a number of generation sources, often in complementary combinations. The goal is to maximize electrical generation based on the resources available while considering their efficiencies and costs. As efficiencies improve, cost savings often result. An electric power plant's *efficiency* (η) is the ratio between the useful electricity output from the generating unit during a specific time period and the energy value of the energy source supplied to the unit [11]. The theoretical efficiency of converting various energy sources into useful electrical energy using a sampling of generation technologies is shown in Table 5.1 [11]. A secondary goal may be to reduce greenhouse gas emissions which is also accomplished by efficiency improvements but may be accomplished by energy conservation, fuel substitution, or using renewables. For example, system electrical generation efficiencies can be improved, and greenhouse gas emissions reduced, by substituting hydropower generation or tidal power for coal or natural gas.

Point of Common Coupling

A microgrid connects to the electric grid at a *point of common coupling* (PCC), the interconnection that maintains microgrid voltage at the same level as the grid unless there is a problem on the grid or other reason to disconnect [9]. The PCC is the location in the electrical system where multiple customers or multiple electrical loads may be connected [12]. It is the point of a power supply network electrically nearest to a particular load at which other loads are connected [13]. These loads can be electrical devices, equipment, or systems, or distinct customer installations [13].

The PCC enables the transfer and exchange of electricity from the microgrid to the larger utility grid. When it is the point of coupling between the microgrid and the host electrical utility, IEEE-519 (a set of guidelines to measure power harmonics on power system connections between a utility line and a building) suggests that this be a point which is accessible to both the utility and the customer for direct measurement [12]. The PCC's primary job is to ensure that voltage regulation from the generation system is synchronized with the voltage from the main power grid. If the grid goes down, it must activate a circuit breaker that isolates the grid. To reconnect, the microgrid requires information about the grid power conditions, such as the frequency and voltage of the grid power. Usually, reconnection is possible within ten seconds or less.

Microgrid Power Management Systems

The power management system handles the transfer of electrical power from the power source to the electricity-consuming devices [14]. These types of electric load management usually require converting the electricity generated from the power source with an inverter that transforms the electricity to the form required for most loads and interfaces with the microgrid's storage components to balance the electrical supply and demand loads [14]. The reliability and maintainability of digital control

TABLE 5.1
Approximate Electrical Generation Efficiencies of Selected Technologies [11]

Renewable generation systems	Efficiency (%)	Fossil fuel generation systems	Efficiency (%)	Other generation systems	Efficiency (%)
Hydro electric		Nuclear Fission	35	Municiple waste	23
Large hydropower	85–98	Natural Gas Turbine	38–42	Sterling engine	38
Small hydropower	25–90	Coal	32–42	Microturbines	20–30
Tidal power	90	Oil	42		
Ocean thermal energy conversion	4	Internal combustion engines			
Wind turbine generators	30–45	Petrol	25		
Solar		Diesel	35–42		
Thermal	25				
Photovoltaic	15–23				
Biomass	25–35				
Geothermal	10–12				
Fuelcells					
Solid oxide	48				
Phosphoric acid	40–45				
Melted carbonate	50				
Proton exchange	40				

system components must be considered as the power management system provides critical services for the operation of the microgrid. The ultimate goal is to create a central microgrid control system that provides active balancing (see Figure 5.3), energy services, and control for energy-consuming devices that are deployed in smart buildings, residential areas, transportation systems, and communities.

The control of the power flow in a microgrid is performed by operation algorithms implemented in the central controller which may utilize weather forecasts and other data [15]. Central controller reference values are distributed to the local power electronic converters whose control loops and topologies are adapted to microgrid requirements [15]. Custom tasks for distributed control systems include stability assurance within the grid and active filtering [15]. Modern microgrid systems often integrate software and control systems, such as smart electric meters, that can manage the grid operation in an efficient and reliable manner [14].

CATEGORIES OF LOADS

Important microgrid components are the electricity loads (consuming devices) whose energy is supplied by the microgrid system [14]. Loads can be divided into groups based on the degree of need for electricity. They are commonly categorized as sensitive, adjustable, or sheddable (see Figure 5.4). Tier 1 loads (sensitive) are those that must operate continuously without fail. These might include elevators, refrigeration equipment, and emergency lighting. When electricity outages occur and impact Tier 1 loads, significant costs, safety issues, or equipment damage are experienced. Tier 2 loads are discretionary (adjustable) and may be shifted or shed for short periods to balance generation availability. Examples include domestic water heating systems, certain fans, and air conditioning loads. Tier 3 loads are those that can be shed for emergency operations due to unplanned and partial loss of generation. Depending on the actual loads, examples might include kitchen equipment, interior lighting, and emergency generation equipment. Some loads may fall into different tier classifications based on the season or time of day.

ENERGY STORAGE SYSTEMS

For many microgrids, batteries are a common type of energy storage technology. They can be designed to convert, store, manage, and recycle energy for extended periods of time.

There are microgrid applications for energy storage technologies which provide brief bursts of power. Ultracapacitor energy storage delivers power services with short response time (cycles ranging from milliseconds to minutes in duration) [7]. When combined with other storage technology, storage times can be extended to several hours. Ultracapacitors are high peak current devices with 100% depth of discharge capability, which provides a reduction in the charge–discharge rate on the battery and lead to longer battery life [7]. Flywheels can provide bursts of electricity for periods of five to 30 minutes.

To ensure the lifecycle and safe operation of the batteries, more complex energy storage systems are equipped with a battery management system (BMS) to monitor

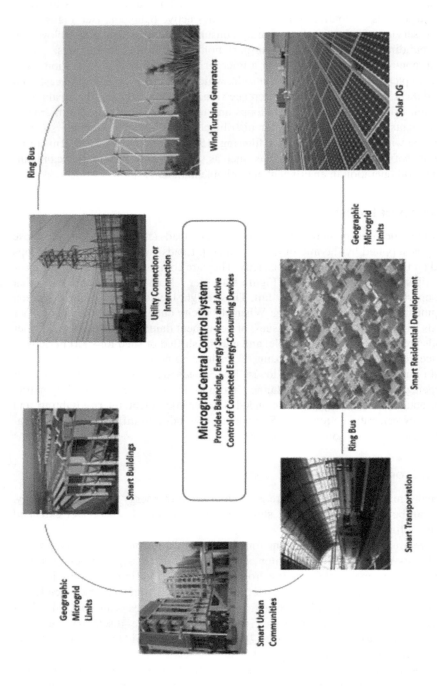

FIGURE 5.3 Microgrid control system to provide active balancing.

Microgrid Architecture and Regulation

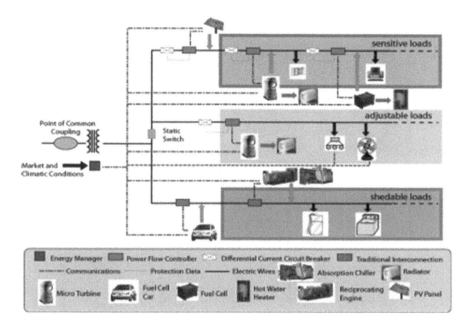

FIGURE 5.4 Graphic showing point of common coupling and types of loads. (Source: Berkeley Lab [16].)

and control the charge and discharge processes of the battery's cells or modules [17]. The configuration for the internal control architecture of a basic BMS would include direct connections from the BMS to the storage controller and the batteries with both also connected directly to a converter. To be fully functional, this configuration further requires an electrical connection and communication to external generation and systems. The control scheme should in turn be determined by the application, which establishes the algorithmic and input/output requirements for the electrical energy storage system [17]. The battery storage architecture for hybrid generation systems can be configured to have an individual BMS for each generation type or share a central BMS and storage system (see Figure 5.5).

ADVANCED MICROGRIDS

Advanced microgrids are a subcategory of microgrids with defined characteristics, some similar to conventional definitions of microgrids with enhanced capabilities that enable the microgrid to offer improved services. According to researchers at Sandia National Laboratories [18], the defining characteristics and features of an advanced microgrid include:

1) being geographically delimited or enclosed
2) having a connection to the main utility grid at one PCC

82 Fundamentals of Microgrids

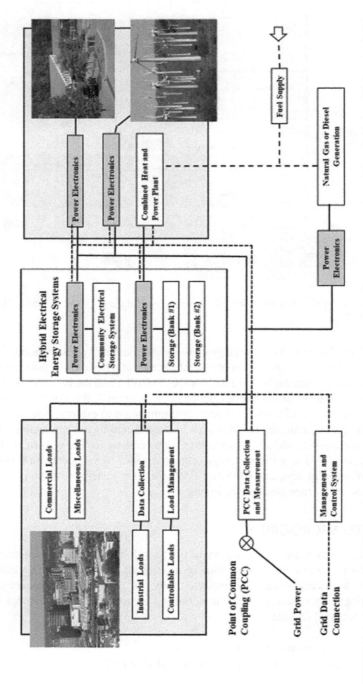

FIGURE 5.5 Microgrid architecture for small to medium size enterprise systems. (Source: concept adapted from Paderborn University [15].)

Microgrid Architecture and Regulation 83

3) being fed from a single substation
4) capability to automatically transit to and from the grid and to be operated in island mode
   - operates in a synchronized and/or current-sourced mode when utility-interconnected
   - is compatible with system protection devices and coordination
5) inclusion of DER while remaining generator-agnostic and in accord with customer needs regarding renewables (inverter-interfaced), as well as having fossil fuel-based (rotating equipment generators) and/or integrated energy storage
6) inclusion of an energy management system (EMS) with controls for power exchanges, generation, load, storage, and demand response, and load-management controls to quickly balance supply and demand
7) inclusion of real-time and instantaneous power and information exchanges across the PCC [18]

Meeting these requirements directly impacts microgrid system architecture by integrating new controls, energy storage, and renewable energy systems to enable greater cost savings and environmental benefits. They achieve plug-and-play interoperability with the use of sophisticated technologies and digital controls that enable peer-to-peer, autonomous coordination among micro-sources [19]. Advanced microgrids typically use inverters and controllers to interface with the EMS or other coupled microgrids [18]. Inverters can provide multiple functions to enable smart-grid interoperability. A challenge for microgrid applications is that the technologies used need to address the optimal mix of power flow [18]. Networked hybrid designs for microgrids may prove to be the optimal configuration to maximize efficiency and performance [18]. With such capabilities, microgrid owners have the capability with advanced microgrids to optimally manage system resources to address threats and potential consequences, and respond quickly to changes in priorities [18].

There are other possibilities for advanced microgrid configurations. Microgrids vary in type, scale, kind of assets, generation capacities, internal connections, and services. The benefit of a grid interconnection includes having supplemental power if the microgrid assets are unable to perform as intended or are offline for routine maintenance. What if there is no opportunity for a grid interconnection at a PCC as there is no central grid available? Rather than being configured for a grid interconnection, multiple microgrids in local proximity could interconnect to support each other at a lower-voltage distribution level essentially creating an interconnected mini-grid (see Figure 5.6). When excess capacity is available that might be useful for other nearby or virtual microgrids, the electricity could be redistributed. Asset and resource coordination would be accomplished from a central control center using wireless remote monitoring and control systems. Communications options might include a local area network with a configuration similar to a token ring.

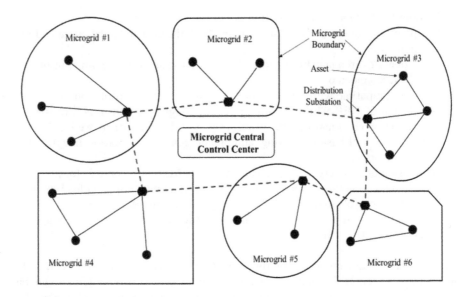

FIGURE 5.6 Architecture for multiple interconnected microgrids coordinating assets.

MICROGRID REGULATIONS AND STANDARDS

Microgrid architecture is strongly influenced by regulations and standards that may vary based on application. Much of the regulation that applies to U.S. microgrid interconnections is created by state governments that are often influenced by established local utility companies rather than the federal government [20]. There is a long history in the electrical utility business of maintaining monopoly status, influencing both regulations and their enforcement procedures. Electric utilities often view non-regulated power generation as a threat which increases their business risks. The paradox is that rather than increasing risks, microgrids when properly designed can actually reduce risks and grid instabilities. Regardless, once regulations are established, maintaining compliance is a chief concern for both public utilities and microgrid operators.

There are understandable concerns about microgrid regulation regardless of the type or configuration of electrical generation systems—especially those that are independent of utility grids. Regulatory policies simply have not been adopted that are congruent with the vision of the utility system provided by the introduction of microgrids. Microgrids, regardless of the architecture employed, encroach on many areas of existing regulation not conceived prior to their recent popularity (i.e., generator interconnection rules, air quality permitting, building codes, tariffs, etc.) [20]. A fundamental example: traditional interconnection rules often require microgrid generators to disconnect when disturbances in the grid occur, though one of the objectives of microgrid development is to achieve systems that can island and fully or partially ride through grid-induced problems [20]. However, the regulatory environment concerning microgrids is in flux and there are a number of legacy recently adopted or updated regulations and standards.

U.S. Clean Air Act

The Clean Air Act is the law that defines U.S. Environmental Protection Agency (EPA) responsibilities for protecting and improving the nation's air quality and the stratospheric ozone layer [21]. The EPA establishes requirements on the use of fossil fuel generation systems which are enforced by state and local governments. Microgrids using fossil fuels must comply with its provisions (see https://www.epa.gov/clean-air-act-overview). Of greatest concern for microgrids is Title I of the Act which covers air pollution and control. Its parts identify requirements for air quality and emission limitations, ozone protection, preventing significant deterioration of air quality, and requirements for nonattainment zones. Title III covers general provisions including administration, licensing, economic impact analysis, and exemptions. Title V identifies permitting requirements.

U.S. greenhouse gas regulations and policies are limited. In April 2009, the EPA released a proposed finding on CO_2 and five additional greenhouse gases (methane, nitrous oxide, hydrofluorocarbons, perfluorocarbons, and sulfur hexafluoride). The agency stated that "in both magnitude and probability, climate change is an enormous problem" and that greenhouse gases "endanger public health and welfare within the meaning of the Clean Air Act" [22]. It cited human-made pollution as a "compelling and overwhelming" cause of global warming. According to EPA administrator Lisa Jackson, the finding "confirms that greenhouse gas (GHG) pollution is a serious problem now and for future generations." On April 2, 2007, the U.S. Supreme Court ruled in *Massachusetts v. EPA* that carbon dioxide could be regulated as a pollutant under the Clean Air Act and that the government had a responsibility to issue a determination based on science. The Bush administration failed to act on this ruling. This finding was prompted by a Supreme Court decision in April 2007 ruling that GHGs are indeed pollutants as classified by the Clean Air Act and regulation is required if human health is threatened [23]. Most federal government programs regarding GHGs are languishing or are being unenforced or dismantled as a result of administrative policy changes beginning in 2016 under the Trump administration. Regardless of the failure of recent federal governmental leadership, individual U.S. states are making impressive policy gains in efforts to reduce GHG emissions (e.g., New York, Connecticut, and California).

U.S. Public Utility Regulatory Policies Act (PURPA)

The PURPA was enacted in 1978 as part of the National Energy Act and amended as part of the Energy Policy Act of 2005. It is the only federal legislation requiring competition in the utility industry. It effected the development of alternative energy and co-generation production by requiring public utilities to purchase power produced by co-generators at reasonable buy-back rates, typically based on the utility company's cost [24]. This effectively created a market for non-utility power producers. PURPA guaranteed that the co-generator or small power producer would be able to interconnect with the electric grid and access backup services from the utility [24]. It also exempted co-generators and small power producers from federal and state utility regulations and the associated reporting requirements of these bodies [24]. The portion of the act dealing with co-generation and small power production appears in U.S. code in Title 16 (Conservation), chapter 12 (Federal Regulation and Development of Power), subchapter II (Regulation of Electric Utility Companies Engaged in Interstate Commerce), section 824a-3: Cogeneration and Small Power Production [25].

To assure the benefits of PURPA, a co-generation facility had to be classified as a qualifying facility, generating electricity and useful thermal energy from a single fuel source [24]. In addition, a qualified co-generation facility had to be less than 50% owned by an electric utility or an electric utility holding company [24]. Finally, the plant had to meet the minimum annual operating efficiency standards established by the Federal Energy Regulatory Commission (FERC) when using oil or natural gas as the principal fuel source [24]. The minimum efficiency standard established was that the useful electric power output, plus one half of the useful thermal output of the facility, must be either: 1) no less than 42.5% of the total oil or natural gas energy input; or 2) 45% if the useful thermal energy is less than 15% of the total energy output of the plant [24].

PURPA is becoming less important as many older supply contracts expire and electric deregulation and open access to electricity transportation by utilities has created a vast market for the purchase of energy [25]. Though PURPA was amended under the Energy Policy Act of 2005 to allow the Federal Energy Regulatory Commission (FERC) to exempt utilities from its requirements if qualifying facilities are provided access to wholesale markets, it has not been updated to reflect the advent of microgrids and need for more competitive markets [26]. In fact, many state regulatory agencies have stopped requiring utilities to offer contracts to developers of non-utility power projects [25]. Regardless, PURPA remains important as it promotes renewable energy (especially hydropower) by exempting the developers of such projects from onerous state and federal regulatory regimes [25].

In 2018, the National Association of Regulatory Utility Commissioners (NARUC) promoted the updating of PURPA for the energy sector [25]. NARUC is proposing that the FERC "exempt from PURPA's mandatory purchase obligation those utilities which are subject to state competitive solicitation requirements and other best practices that ensure all technologies access to the market" [26]. Citing examples that qualifying facilities have invoked PURPA in an anti-competitive manner, NARUC suggested that reforms would better enable competitive solicitation and procurement [26].

IEEE Standard 1547-2018 for Interconnection and Interoperability of Distributed Energy Resources with Associated Electric Power Systems Interfaces

Institute of Electrical and Electronics Engineers (IEEE) Standard 1547-2018 outlines the requirements and technical specifications for interconnecting distributed resources safely with the grid. This includes design, safety, response to abnormal conditions, power quality, equipment, production, installation, commissioning, and periodic testing [27]. This standard was first published in 2003, after five years of development, with the goal of creating a set of technical requirements that could be used by all parties on a national basis [28]. It also addresses unintentional islanding by requiring detection of the island condition and the ability to cease energizing the area electric power system (EPS) within two seconds of the microgrid's island formation [28]. It emphasizes installation of DER on radial primary and secondary distribution systems using 60 Hz sources. IEEE 1547 also provides anti-islanding standards to protect the safety of utility line workers [2].

IEEE P1547.4-2011 Guide for Design, Operation, and Integration of Distributed Resource Island Systems with Electric Power Systems

Intended for use by designers, operators, system integrators, and equipment manufacturers, IEEE P1547.4 covers microgrids and intentional islands that contain distributed resources connected at a facility and with the local utility [28]. The guide covers the distributed resource, interconnection systems, and participating electric power systems [28]. It provides alternative approaches and good practices for the design, operation, and integration of microgrids and covers the ability to separate from and reconnect to the utility grid while providing power to the islanded local power systems [28]. It is relevant to the design, operation, and integration of distributed resource island systems [29]. Implementation expands the benefits of using distributed resources by targeting improved reliability and builds upon the interconnection requirements of IEEE Standard 1547(TM)-2008 [29].

IEEE P2030.7-2007 IEEE Standard for the Specification of Microgrid Controllers

Microgrid controllers are direct digital control (DDC) systems that are responsible for managing the operating equipment within the microgrid. A key element of microgrid operation is the microgrid energy management system (MEMS). IEEE P2030.7-2007 is an international standard that defines how microgrid control functions operate as self-managing, autonomously, or grid-connected, and seamlessly connect to and disconnect from the main distribution grid for the exchange of power and the supply of ancillary services [30]. The standard categorizes microgrid functional control assignments into four blocks: grid-interactive control functions, supervisory control functions, local area control functions, and device level control functions [31].

The scope of this standard is to address the functions above the component control level associated with the proper operation of the MEMS that are common to all microgrids, regardless of topology, configuration, or jurisdiction [30]. It also describes control approaches required from the distribution system operator and the microgrid operator [31]. IEEE Standard P2030.7-20078 addresses testing procedures and attempts to ensure interoperability of the microgrid by defining which parts of the microgrid controller must be standardized and which can remain proprietary [27].

IEC 61727 International Electrotechnical Commission's PV System Requirements

This international standard applies to utility-interconnected photovoltaic (PV) power systems operating in parallel with the utility and utilizing static (solid-state) non-islanding inverters for the conversion of DC to AC power. It describes the international requirements for photovoltaic systems interconnecting with existing or proposed low-voltage utility distribution systems. Recent updates provide clarification regarding for non-islanding inverters, point of common coupling, and power factor.

MICROGRID STANDARDS BEING DEVELOPED [25]

IEC TS 62898-1: Guidelines for Microgrid Projects Planning and Specification; published

IEC TS 62898-2: Microgrids Guidelines for Operation; working document

IEC TS 62898-3-1 Microgrids: Technical/Protection Requirements; working document

IEEE P1547 REV: Microgrid Connection to Distribution Utilities; Microgrid/Distribution Utility, ISO/RTO

IEEE P157.8/D5.0: Draft Recommended Practice for Establishing Methods and Procedures that Provide Supplemental Support for Implementation Strategies for Expanded Use of IEEE 1547 (Clause 8, Recommended Practice for DR Islanded Systems)

IEEE P2030.10: Standard for DC Microgrids for Rural and Remote Electricity Access Applications

IEEE P2030.9: Recommended Practice for the Planning and Design of the Microgrid

IEEE P2030.7: Standard for the Specification of Microgrid Controllers

IEEE P2030.8: Standard for the Testing of Microgrid Controllers

SUMMARY

Microgrids are normally configured as either decentralized or distributed networks. While there are basic components fundamental to the architecture of a microgrid, the microgrid must contain the equipment necessary to generate electricity. They may also have a management system, loads, and possibly storage capabilities. Microgrids can be designed to be completely grid-independent or connected and capable of operating in island mode. Unlike traditional electrical distribution systems, microgrids are capable of providing bidirectional transfer of electricity, either from the host grid to the microgrid or from the microgrid to the grid. They can be designed to use AC or DC power or both.

A feature of those capable of islanding is the point of common interconnection between the microgrid and its host grid. Advanced microgrids are a subcategory of microgrids with defined characteristics, some similar to conventional microgrids but having supplemental capabilities that enable them to provide improved services for their customers. Toward the goals of maximizing electrical generation based on resource availability, production efficiencies, and costs, microgrid operators can select from a number of complementary combinations of generation sources. Microgrid control systems strive to achieve active balancing and may categorize connected loads as sensitive, adjustable, or sheddable. They can also be interconnected and rely on other microgrids for power rather than a central grid.

Microgrid architecture is influenced by the regulations that apply to their development. Those using fossil fuels must comply with the provisions of the U.S. Clean Air Act, primarily Title I of the Act which sets standards for air pollution. The IEEE has established standards for microgrid interconnection requirements and provides microgrid design, operation, and interconnection guidelines. Other microgrid standards are evolving. Microgrids create a special regulatory issue as many U.S. states and local governments have not yet adopted regulations for their design and implementation. However, many have regulations concerning the development of distributed energy resources and guidelines concerning utility interconnections.

ACKNOWLEDGMENTS

This chapter is a revised version of a reviewed article entitled "Microgrid Architecture" that was originally published in the *International Journal of Strategic Energy and Environment Planning,* Volume 2, Issue 5, Stephen A. Roosa, Ph.D. (author/editor) in September 2020. The editor extends thanks and credit to the Association of Energy Engineers, Atlanta, Georgia for use of previously published material and for approving that the article be republished.

REFERENCES

1. Eger, K., Goetz, J., Sauerwein, R., Frank, R., Boëda, D., Martin, I., Artych, R., Leukokilos, E., Nikolaou, N. and Besson, L. (2013). Microgrid functional architecture description. FI.ICT-2011-285135 FINSENY D3.3 v1.0, page 3.
2. Microgrid Institute (2014). About microgrids. http://www.microgridinstitute.org/about-microgrids.html, accessed 11 November 2018.
3. Lumbley, M. (2018, November). Beyond politics, a project manager's argument for clean energy. *North American Clean Energy,* 12(6), page 64.
4. S&C Electric Company (2018). Is a microgrid right for you? Part 1 of 3. https://www.sandc.com/globalassets/sac-electric/documents/sharepoint/documents---all-documents/education-material-180-4504.pdf, accessed 24 January 2020.
5. Eger, K., Goetz, J., Sauerwein, R., Frank, R., Boëda, D., Martin, I., Artych, R., Leukokilos, E., Nikolaou, N. and Besson, L. (2013, March 31). Microgrid functional architecture description. FI.ICT-2011-285135 FINSENY, D3.3 v1.0.
6. Blackhaus, S., Swift, G., Chatzivasileiadis, S., Tschudi, W., Glover, S., Starke, M., Wang, J., Yue, M. and Hammerstrom, D. (2015, March). DC microgrids scoping study—estimate of technical and economic benefits. Los Alamos National Laboratory. https://www.energy.gov/sites/prod/files/2015/03/f20/DC_Microgrid_Scoping_Study_LosAlamos-Mar2015.pdf, accessed 16 January 2019.
7. Maxwell Technologies, Inc. (2018). Microgrids. https://www.maxwell.com/solutions/power-grid/microgrids, accessed 28 November 2018.
8. Joos, G. (2013, November 4). Integration and interconnection of distributed energy resources. Presentation, University of Illinois Urbana-Champaign.
9. Lantero, A. (2014, June 17). How microgrids work. U.S. Department of Energy. https://www.energy.gov/articles/how-microgrids-work, accessed 24 January 2020.
10. Schutte, S. (2018). The power next door. Centrica. https://www.centrica.com/platform/the-power-plant-next-door?fbclid=IwAR2jw2HGv_XVbP0GRP7IFddsjLTPAodhHsurQXjzaPdFWlTntsvBJWtkCBI, accessed 24 January 2020.

11. Electropaedia (Woodbank Communications, Ltd.). Battery and energy technologies, energy efficiency. https://www.mpoweruk.com/energy_efficiency.htm, accessed 2 December 2019.
12. Nabi, M. and Anusha, S. (2016). Voltage sag and support in power control on distributed generation inverters. *International Journal for Modern Trends in Science and Technology*, 2(1), pages 193–198.
13. Encyclo.co.uk. Point of common coupling. http://www.encyclo.co.uk/meaning-of-point%20of%20common%20coupling, accessed 5 February 2020.
14. Miret, S. (2015, February 25). How to build a microgrid. http://blogs.berkeley.edu/2015/02/25/how-to-build-a-microgrid, accessed 26 November 2018.
15. Paderborn University. Power electronics and electrical drives. https://ei.uni-paderborn.de/en/lea/research/forschungsprojekte/power-electronics/sme-microgrid-with-intelligent-power-controllers, accessed 28 November 2018.
16. Berkeley Lab (2018). About microgrids. https://building-microgrid.lbl.gov/about-microgrids, accessed 26 November 2018.
17. International Electrotechnical Commission (2011). Electrical energy storage, white paper. http://www.iec.ch/whitepaper/pdf/iecWP-energystorage-LRen.pdf, accessed 26 January 2020.
18. Sandia National Laboratories (2014, March). The advanced microgrid: integration and interoperability. Sandia Report SAND2014–1535.
19. Eto, J. (2015, October 8). The CERTS microgrid. Brookhaven National Laboratory smart grid workshop. https://www.bnl.gov/sgw2015/files/talks/Eto.pdf, accessed 17 May 2020.
20. Marnay, C., Asano, H., Papathanassiou, S. and Strbac, G. (2008, May/June). Policy making for microgrids. *IEEE Power and Energy Magazine*, page 69.
21. U.S. Environmental Protection Agency. The Clean Air Act. https://www.epa.gov/clean-air-act-overview/clean-air-act-text, accessed 4 February 2020.
22. POWERnews (2009, April 22). EPA finds greenhouse gases pose threat to public health, welfare. https://www.powermag.com/epa-finds-greenhouse-gases-pose-threat-to-public-health-welfare, accessed 12 January 2019.
23. Shin, L. (2009, February 18). EPA review reverberates through U.S. energy industry. http://solveclimate.com/blog/20090218/epa-review-reverberates-through-u-s-energy-industry/, accessed 29 April 2009.
24. Thumann, A. and Mehta, P. (2018). Standards, codes, and regulations, Chapter 20, in *Energy Management Handbook* (Roosa, S., ed.). Fairmont Press: Lilburn, Georgia, pages 549–550.
25. Wikipedia. Public Utility Regulatory Policies Act. https://en.wikipedia.org/wiki/Public_Utility_Regulatory_Policies_Act, accessed 9 December 2018.
26. Brandt, J. (2018, October 22). NARUC urges FERC to support expanding competitive practices under PURPA. *Daily Energy Insider*.
27. S&C Electric Company (2018). How to build a microgrid, Part 2 of 3. https://www.sandc.com/globalassets/sac-electric/documents/sharepoint/documents---all-documents/education-material-180-4505.pdf, accessed 24 January 2020.
28. Marney, C., Asano, H., Papathanassiou, S. and Strbac, G. (2008, May). Policymaking for microgrids. *IEEE Power and Energy Magazine*, page 75.
29. IEEE Xplore Digital Library. https://ieeexplore.ieee.org/document/5960751.
30. IEEE Standards Association. IEEE P2030.7-2007 – IEEE standard for the specification of microgrid controllers. https://standards.ieee.org/standard/2030_7-2017.html, accessed 9 December 2018.
31. Reilly, J., Hefner, A., Marchionini, B. and Joos, G. (2017, October 24). Microgrid controller standardization – approach, benefits and implementation. *Conference Presentation at Grid of the Future*. Cleveland, Ohio.

6 Linking Microgrids with Renewable Generation

THE IMPACT OF RENEWABLE ENERGY

Alternative and renewable energy refer to electricity or heat generated from renewable sources (including wind, solar, geothermal, biomass, landfill gas, and low-impact hydro) as well as energy substitutes and various conservation methods. Renewable energy is considered to be more environment-friendly than traditional electricity generation. Unlike fossil fuels, alternative and renewable energy emit little or no air pollution. They also do not generate radioactive wastes. Most importantly, they are replenished by naturally occurring processes. Investments in renewable energy have been increasing and the economic impact is considerable. Worldwide, annual renewable power generation investments increased to $148 billion by 2007 and to $242 billion by 2014, greater than that year's $132 billion invested in fossil fuel systems [1]. Clean energy investments (renewables plus nuclear energy) exceeded $288 billion in 2018.

In contrast to green power or electricity generated from renewable sources, *brown power* is electricity generated using environmentally hostile technologies. The vast majority of electricity in the United States comes from the combustion of coal and natural gas. The remaining amounts of electrical energy are produced mostly from nuclear power, hydropower, wind, solar, and other portable fossil fuel sources. The brown power generators are sources of air pollution in the U.S. and elsewhere and contribute to smog and acid rain. They are also the greatest contributors of greenhouse gasses (GHGs), including CO_2 and nitrous oxide. Innovative alternate energy technologies can decrease the reliance on imported oil and fossil fuels. The combustion of oil and its byproducts contribute to GHG emissions but are used primarily in vehicular transportation.

The use of alternative forms of energy, including wind and solar power, began increasing in the U.S. with wind and solar energy output jumping roughly 45% from 2006 to 2007. The U.S. share of renewables to total domestic electrical use doubled between 2010 to 2017 (see Figure 6.1). This was driven by steady reductions in the costs of the technologies, tax credits, and other incentives and policies that focused on sustainable energy development. U.S. renewable energy sources—including biomass, geothermal, hydropower, solar, and wind—accounted for 17.8% of the country's net domestic electrical generation through the first nine months of 2018 [2]. Through the same period, Energy Information Administration (EIA) data shows that solar and wind both saw strong growth, with utility-scale solar growing by 30.3% (which includes distributed small-scale solar) and wind energy growing by 14.5% compared to the same period a year earlier [2]. Together, wind and solar accounted for 9% of the country's electrical generation and almost 50% of

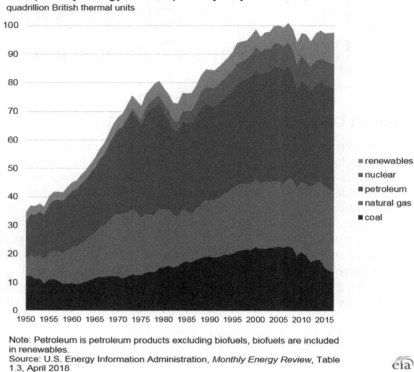

FIGURE 6.1 U.S. Primary energy consumption by resource (EIA).

the total from all renewable energy sources [3]. For renewable energy technologies, hydropower accounted for 7.05% of the nation's total electricity generation, followed by wind with 6.41%, solar with 2.42%, biomass with 1.48%, and geothermal with 0.39% [3].

There are changes in the mix of energy being used in the U.S. There is an obvious trend away from coal as a primary combustion fuel to natural gas and renewables. Capacities for renewable generation exceed that of coal in the U.S. as does production. According to the EIA, in April of 2019 renewables provided 68.5 million MWh of electricity compared to just 60 million MWh from coal. It is likely that 100–150 natural gas-fired plants will be constructed in the U.S. within the next ten years. However, the greater use of natural gas is not compatible with the ultimate goals being set by U.S. cities and localities to substantially reduce greenhouse gas emissions that target reductions approaching 100%. While these goals are in some cases idealistic, they are nevertheless notable. With a view of sustainably developing future electrical power generation and transmission, more efficient technologies are needed [4]. Many municipal planning agencies intend to rely on renewables, particularly solar and wind power, to meet their GHG reduction goals. While hydropower production has been stable, much of the growth in renewable energy sources can be attributed to increased use of wind power and to a lesser extent biomass and solar. These trends are creating new markets for the development of microgrids that use renewable energy.

Worldwide, a 20% to 25% increase in renewable energy production should cause a considerable reduction in CO_2 emissions, if the use of fossil fuels is stabilized or reduced. China ranks first internationally in installed renewable generation including hydropower, wind power, and solar energy. It plans for 50% of its power generation by 2030 to be from renewable energy sources which will require an investment of more than 10 trillion yuan ($1.4 trillion U.S.) [5]. To meet this target, installed non-fossil fuel power generation capacity will need to increase by one billion kW, an amount equaling the current total installed generation capacity of the U.S. [5].

RENEWABLE GENERATION FOR MICROGRIDS

A guiding design principle for buildings and energy-consuming systems has been to initially focus on energy efficiency and conservation; then, after these options are exhausted and optimized, focus on renewable energy to supply the needed power [6]. In conjunction with efforts toward energy efficiency and associated conventional technologies, renewable generation can be developed to effectively reduce GHG and CO_2 emissions in a sustainable manner, sometimes with substantial cost savings. The inherent inefficiencies in the conventional electric grid present a remarkable opportunity for microgrids: reductions in on-site electricity consumption can result in a three-fold reduction in non-renewable energy use, thus providing opportunities for smaller amounts of renewable generation to locally supply the remaining power required [6]. This leveraging impact on system energy consumption when renewable microgrids are deployed is often not cited in the literature.

Certain types of renewable energy, such as hydropower, geothermal energy, and in some cases biomass, can provide microgrids with baseload electrical generation.

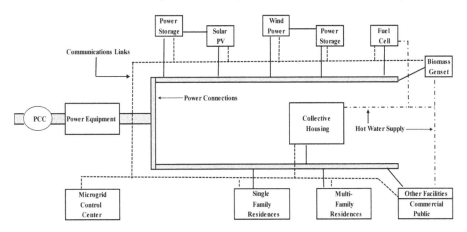

FIGURE 6.2 Microgrid configuration with renewable generation, energy storage, and district heating. (Source: adapted from reference [4].)

Others, such as solar and wind power, provide microgrids with variable renewable energy (VRE) and require batteries or some other form of energy storage to increase capacities (see Figure 6.2). For electrical generation systems, *capacity* refers to the percentage of time over a given period (typically a year) that the system is capable of producing electricity.

SMALL HYDROPOWER SYSTEMS

Hydropower is a key means of generating electricity and offers a set of highly scalable applications for the generation of electricity. In 2017 hydropower accounted for 43% of the renewable electricity generated and 25% of the total renewable energy consumed in the U.S. [7]. However, most of this generation is provided by large-scale, high-head hydropower projects. Very little is provided by small-scale hydropower systems that support microgrids. This is changing somewhat as improvements in small-scale technologies that can support microgrid development are being adopted and deployed.

Of the total solar energy incident on Earth, approximately 21% is used for maintaining the global water cycle of evaporation and precipitation. But only 0.02% of this amount of energy is available as kinetic and potential energy stored in the rivers and lakes of the Earth [8]. The water reserves are available in solid (ice), liquid (water), and gaseous (water vapor) conditions and are continuously cycled by incident solar energy. This global water cycle is mainly fed by evaporation of water from oceans, plants, and continental waters. The resulting precipitation which feeds snow fields, glaciers, streams, rivers, lakes, and groundwater is a source for small hydropower. Using low-head hydropower systems or flowing water (run of the river) to power electric turbines creates opportunities to generate power in close proximity to loads as many villages and towns are found near rivers and streams. Small hydropower plants are less than 30 MW in size. There are thousands of such sites in the U.S. with existing dams that are available for mini-hydropower applications. A small hydropower installation at the Kentucky River Lock and Dam No. 2 in Mercer County, Kentucky (the Mother Ann Lee hydroelectric plant) began operation in 2007 and provides power for the equivalent of 2,000 residences [9]. There are numerous recent international examples including the new 15 MW hydropower plant in the Soloman Islands.

Non-traditional forms of hydropower include technologies and structures that harness the power of water in tides, waves, and temperature differences. Tidal power systems include tidal fences and barrages that enable the extraction of tidal energy by constructing a barrage across an estuary or bay to produce static head. Wave systems that generate electricity include permanent magnet linear generator buoys that are anchored to the seafloor, devices that ride on the waves, and capture chambers that use waves to pressurize air. Ocean thermal energy conversion (OTEC) systems take advantage of thermoclines in tropical regions and use a condenser/compressor cycle to drive a turbine. The closed cycle OTEC system is similar to a conventional Rankine cycle steam engine. It is essentially the same as commercial refrigeration, except it runs in the opposite direction such that a heat source and a cold sink are used to produce electric power rather than the reverse. OTEC uses a different working fluid (typically ammonia) and operates at low temperatures and pressures. Due to the large amounts of water that must be pumped, OTEC systems operate at very low efficiencies.

Biomass Energy

Biomass resources release solar energy stored in plants and organic matter by burning agricultural waste and other organic matter to generate electricity including landfill methane-gas-to-energy conversion. Biomass can be divided into primary and secondary products. The former is produced by directly using solar energy through photosynthesis. In terms of energy supply, these are farm and forestry products from energy crop cultivation (e.g., fast-growing trees such as poplars or grasses such as switchgrass), plant byproducts, residues, and waste from farming and forestry operations including the downstream industries and households (e.g., straw, residual and demolition wood, or organic components in household and industrial waste). Secondary products are generated by the decomposition or conversion of organic substances in higher organisms (e.g., the digestion system of animals); examples include liquid manure and sewage sludge.

Biofuels are used for electrical generation in microgrids. The traditional way to generate electricity using primary biomass materials is by direct combustion. The City of Burlington, Vermont uses wood from managed forestry operations to generate baseload electricity for its microgrid. Biomass can also be processed using gasification technologies to create suitable fuels to generate electricity. Ethanol is normally used as a transportation fuel but is corrosive and is difficult to transport long distances in pipelines. An alternative is to use it to generate electricity. Brazil is the world's second largest ethanol producer after the U.S. The city of Juiz de Fora (population 150,000) in Minas Geris generates electricity with ethanol made from sugar cane [10]. The plant, a simple-cycle, natural gas system, uses a converted 43.5 MW combustor and has an operating capacity of 87 MW [10].

Landfill Gas Extraction

A landfill is a waste disposal site. Does this represent a form of renewable energy? The answer is yes to the extent that landfill gas is produced by organic decomposition. At municipal sanitary landfills, wastes are generally spread in thin layers, compacted, and covered and compacted further with fresh layers of soil, clay, or plastic foam each day [11]. The decomposing wastes generate methane (CH_4) gas, a hydrocarbon molecule that is the principal component of natural gas. CH_4 is produced through anaerobic (without oxygen) decomposition of waste in landfills, animal digestion, decomposition of wastes, production and distribution of natural gas and petroleum, coal production, and incomplete fossil fuel combustion. It is also formed and released into the atmosphere by biological processes occurring in anaerobic environments.

Once in the atmosphere, CH_4 absorbs terrestrial infrared radiation that would otherwise escape to space. This property contributes to the warming of the atmosphere, which is why CH_4 is classified as a greenhouse gas. CH_4 has global warming potential estimated to be a minimum of 23 times higher than CO_2. CH_4's chemical lifetime in the atmosphere is approximately ten years but it absorbs much more energy and has indirect effects such as being a precursor to ozone [12]. Methane's relatively short atmospheric life, coupled with its potency as a GHG, makes it a candidate for mitigating global warming over the near term (e.g., the next 25 years or less).

Projects that transform landfill gas (LFG) to energy significantly reduce biomethane emissions from municipal solid waste (MSW) landfills—and simultaneously create green energy [13]. These projects offset the use of fossil fuels such as coal and natural gas. MSW landfills are one of the largest human-generated sources of methane emissions in the U.S., accounting for 14.1% of methane emissions 2017. In general, LFG projects could capture roughly 60% to 90% of the biomethane emitted from the landfill, depending on system design and effectiveness. The captured methane is converted to water and less potent CO_2 as the gas is burned to produce electricity. LFG can be upgraded to renewable natural gas (RNG), a high-Btu gas, through treatment processes by increasing its methane content while reducing its CO_2, nitrogen, and oxygen contents [14]. Producing energy from renewable RNG avoids the need to use non-renewable sources such as coal, oil, or natural gas from traditional sources (Figure 6.3).

Over 70% of the operational LFG projects in the U.S. are used to generate electricity. About 10% of currently operating LFG energy projects use RNG [14]. The technologies used include internal combustion engines, turbines, microturbines, and fuel cells [14]. Reciprocating engines are the most commonly used conversion technology for LFG electricity applications due to their relatively low cost, high efficiency, and sizes that complement the gas output of landfills [14]. Landfill gas

FIGURE 6.3 Genset using landfill gas in Johnson City, Tennessee.

projects operate about 95% of the time, providing the potential to reduce baseload electrical power and greenhouse gases that would normally be flared.

Many companies have developed successful operations across jurisdictions that help achieve the goals of sustainability. Waste Management, a company that operates 281 landfills across the U.S., is turning the production of electricity from renewable landfill gases into a major business opportunity. Landfill gases are typically composed of 40% to 50% CO_2 and 50% to 60% CH_4 [15]. Waste Management currently operates 103 landfill gas-to-energy (LFGTE) plants. The company intends to create new revenue streams by developing additional sites, with another 60 planned adding 230 megawatts to its total generating capacity [15]. These projects will be located in Texas, Virginia, New York, Colorado, Massachusetts, Illinois, and Wisconsin. When completed, the company will become a mini-utility. Paul Pabor, the company's vice-president of renewable energy, sees these initiatives as "a major step in Waste Management's ongoing efforts to implement sustainable business practices" [15]. Combined with their existing LFG generation facilities, the total electrical energy produced when all projects are completed will be 700 megawatts—roughly the equivalent of a fossil fuel power plant [15].

Solar Energy (Thermal)

Solar energy initially arrives at the outer edge of the Earth's atmosphere. Including all types of solar radiation, it is measured by satellites as being equivalent to an average of 1.3615 kW/m². This value is referred to as the *solar constant* as it has varied only 0.2% in the last 400 years. Part of this radiation is reflected back to outer space, part is absorbed by the atmosphere and re-emitted, part is scattered by atmospheric particles, and part is absorbed by the Earth's surface and radiates into the atmosphere [6]. Only about two-thirds of the sun's energy reaches the surface of the Earth. Peak solar radiation on the surface of the Earth is approximately 1 kW/m². Solar radiation (*insolation* or *irradiance*) data are often given in kWh/m²/day, sometimes referred to as *peak sun hours*, a term for the solar insolation that a particular location would receive if the sun were shining at its maximum value for a certain number of hours [6]. The types of applications used to generate electricity include solar thermal and solar photovoltaic (PV) technologies.

Solar thermal (also called thermosolar) applications use solar energy for water heating, space heating, or electrical generation applications. They typically employ mechanical systems to collect, distribute, and/or store solar-heated fluid or air [6]. In industrial systems, energy demand will rarely correlate with solar energy availability. In some cases, the energy can be stored until needed, but in most systems, there will be some available solar energy that will not be collected. Successful solar-thermal applications are a function of design, site conditions, climate, and project costs. These systems are usually sized in terms of collector panel area. Pumps, piping, heat exchangers, and storage tanks are then selected to match [6].

A wide variety of devices may be used to collect solar energy. *High-temperature collectors* (>50°C above ambient temperature), such as evacuated tubes, are used mainly for industrial process heat, solar air conditioning, and (rarely) water heating in buildings [6]. Flat-plate collectors are most common in building applications while concentrating collectors are more likely to be used for electrical generation. Technologies that use concentrating solar power (CSP) technologies to harness solar thermal energy for electrical generation include parabolic troughs, central receivers, dish-Stirling engines, linear Fresnel reflectors, and power towers. To increase the capacity of solar thermal generation, they are often coupled with energy storage systems.

Solar Energy (Photovoltaic)

Solar photovoltaic is today a very common way to generate distributed electricity. Manufactured using mass production technologies, solar panels can be installed almost anywhere and will generate power as long as the sun is shining. Due to their flexibility, ease of transport, installation alternatives, and modularity solar arrays are a preferred means of providing distributed generation, especially in remote locations where electrical costs are higher (see example shown in Figure 6.4). The primary types of commercially available solar PV technologies are thin film, polycrystalline silicon, and monocrystalline silicon. In most applications, the largest expenses are the solar collector panels (modules), support structures (mounting and racking systems), electrical connections, and labor [6].

As of mid-2018, there were more than 2,450 utility-scale solar photovoltaic sites in the U.S. representing about 27.5 gigawatts (GW) of installed capacity [16]. Although more than half of installed PV capacity comes from sites 50 MW or larger, most of the total capacity and utility-scale PV facilities are 5 MW or less [16]. Small utility-scale facilities (ranging from 1 MW through 5 MW) account for 18% of the total installed solar capacity and 72% of utility-scale solar facilities located in all 50 states [16].

FIGURE 6.4 Solar PV installations on the island of Antiqua (2018).

Community-developed solar facilities, averaging 3 MW, offer a share of their solar capacity for sale to off-site customers who may not necessarily have access to solar generation [16]. In these programs, customers subscribe to a designated community solar facility and receive monthly credits on their electric bills for the energy generated based on the share of solar capacity purchased [16]. An example of the role of community solar has in the growth of small utility-scale solar facilities is in the U.S. state of Minnesota where community solar facilities account for more than 36% of the total operational PV capacity [16].

Solar PV systems use combinations of panels mounted in arrays. Racking systems use either fixed-tilt systems that are rigidly set at an angle, or tracking systems that follow the movement of the sun's path in the sky. Fixed tilt systems are the least costly.

The maximum energy collected by the solar PV modules occurs when the sun's rays are perpendicular to the panel. The amount of electricity generated by a fixed-tilt solar PV system depends on the orientation of the PV panels relative to the sun [17]. Almost 40%, or 10.4 GW, of utility-scale solar PV systems installed in the U.S. at the end of 2017 were fixed-tilt arrays [17]. Of the utility-scale fixed-tilt solar PV systems, 76% of the capacity was installed at a fixed angle between 20° and 30° from the horizon [17]. For U.S. installations the angle of the tilt is related to the latitude of the site. Systems installed with tilt angles less than the latitude angle of the plant's location typically have greater output during summer months when the mid-day sun is higher in the sky and reduced output during winter when the sun tracks lower in the sky [17]. In the U.S., most of the output from solar PV installations is from sites located between 30° to 40° north latitude (see Figure 6.5).

Solar PV is a preferred solution for microgrids. The Moapa Southern Paiute Solar Project is a 250 MW microgrid that uses solar PV to generate enough electricity (AC) to power an estimated 111,000 homes [18]. Located on the Moapa River Indian reservation approximately 30 miles (48 km) from Las Vegas, this facility is the first

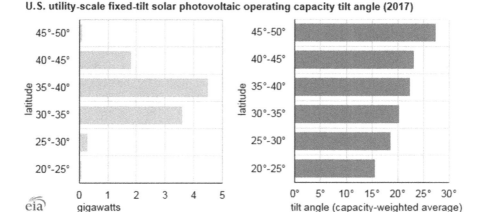

FIGURE 6.5 U.S. utility-scale fixed-tile solar PV capacities and tilt angle. (Source: U.S. Energy Information Administration, Annual Electric Generator report.)

utility-scale solar power plant to be built on tribal lands. It has a 25-year power purchase agreement (PPA) with the Los Angeles Department of Water and Power to supply renewable electricity [18]. Power is generated using 3.2 million (25 million ft^2, 2.32 million m^2) advanced First Solar thin film photovoltaic solar panels [18].

WIND POWER

Wind power can be harnessed to provide mechanical power or to generate electricity. Windmills are used for applications such as water pumping. Wind turbine generators (WTGs) convert the kinetic energy contained in the wind first into mechanical energy and next into electricity by turning a generator [7]. WTGs can be used to support microgrid applications. The size of the turbine and the speed of the wind are key determinates in how much electricity can be produced [19]. A small wind energy system will produce a power output of 100 KW or less [19]. Small commercial and community wind projects typically use turbines with ratings from 100 KW to 1.5 MW. Utility-scale WTGs typically range from 1.5 MW to 8 MW. A 12 MW offshore WTG prototype, the world's largest, is planned for installation and testing in The Netherlands in 2019 [20]. The Haliade-X turbine is designed to generate 67 GWh annually with a 63% capacity factor [20]. Specifications include a 220 m rotor, 107 m blades, and features which make the model less sensitive to wind speed variations [20].

There are two primary types of WTGs, vertical axis and horizontal axis. Modern horizontal axis WTGs with three blades that face into the wind are the most common configuration [7]. Wind speeds are a determining factor in assessing a WTG's performance since faster wind speeds contain more kinetic energy. Generally, annual average wind speeds of greater than 9 miles/hour (4 m/s), are required for small electric wind turbines not connected to the grid; utility-scale wind plants require a minimum wind speed of 10 miles/hour (4.5 m/s) [7]. The power available to drive wind turbines is proportional to the cube of the speed of the wind; therefore, doubling the wind speed leads to an eight-fold increase in power output [7].

The wind farm at Tehachapi Pass near Mojave, California (see Figure 6.6) is one of the world's largest. The site has over 5,600 WTGs with a total annual production over 800 million kWh annually. Initially established in the early 1980s without a utility grid-connection, the Tehachapi wind resource area now exports electrical energy to other parts of California. This was accomplished by the completion of the Tehachapi Renewable Transmission Project in 2012 that connects transmission to the grid serving Los Angeles.

GEOTHERMAL ENERGY

The more ubiquitous uses of geothermal (Earth heat) energy resources are grouped into the categories of direct use and geothermal heat pump applications. Less common is the use of the Earth's subsurface heat to generate electricity. Basically, the process involves extracting the heat from the Earth's molten core and using it to drive a turbine and generator to generate electricity. This core consists primarily of extremely high temperature liquid rock known as magma. The heat circulating

Linking Microgrids with Renewable Generation 101

FIGURE 6.6 Wind turbine generators at Tehachapi Pass, California.

within the rock is transferred to underground reservoirs of water which circulate under the Earth's crust. The geothermal resources tapped to generate electricity are intense and can reside as deep as 10,000 feet (about 3,050 m) below the Earth's surface [21]. The closer the resource to the Earth's surface, the less the costs of drilling.

The basic process of extracting geothermal energy to generate electricity starts with drilling a hole in the ground to release the naturally heated water. The high-temperature water remains in a liquid state as it is under pressure. It changes state, becoming steam when it reaches the Earth's surface and is exposed to atmospheric pressure. This steam is used to drive a turbine. After its heat is neutralized by exposure to lower temperatures, the steam changes back into water. This water is then reinjected to replenish the sub-surface reservoir which is again reheated in a continuous cycle. The key variables when estimating the size of the resource include field size, depth below ground level, and water pressure. Testing helps to evaluate potentials and how quickly and in what volumes a field can heat and release water. It also provides information that helps manage field resources and prevents rapid resource depletion. While this is rare, several wells in California have been retired that were developed as early as the 1970s due to resource depletion.

The goal of the process is to generate electricity by creating a continuous heat transfer system. There are several technologies that enable geothermal energy to be used for electrical generation. Geothermal electricity generation technologies consist of either *flash* technology or *binary* technology or a combination of the two. With flash technology, water from 300°F to 700°F, still in a liquid form, is piped from its highly pressurized underground reservoir into a geothermal facility [21]. Once this super-heated water is released, it flashes into steam that then drives a conventional turbine generator [21]. Using binary technology, underground reservoirs of lower temperature water are used for flashing [21]. Binary technologies use such water to transfer the heat via a heat exchanger to a secondary liquid (isobutane or pentafluoropropane) that changes into a steam-like, gaseous vapor to drive turbines to generate electricity [21]. Dry steam, a rarer resource, is used for larger geothermal plants such as those in northern California which hosts the world's largest number of geothermal plants. When available, highly pressurized geothermal dry steam can be used to directly drive a turbine.

Advances in generation technologies have enabled engineers to use water at lower temperatures. New drilling technologies borrowed from the oil and gas industry, such as directional and horizontal drilling, have enabled the use of well fields that were previously difficult to access. Geothermal resources can be accessed by using more expensive technologies such as enhanced geothermal. Using this process, hydraulic stimulation or shearing creates fractures in crystalline bedrock for applications such as small-scale generation of district heating.

Geothermal energy is a match for microgrids. The U.S. maintains world leadership in generating electricity with about 3,600 MW of capacity. Geothermal electrical generation technologies have been tested and proven reliable. They eliminate most of the combustion cycle problems associated with fossil fuels. There are also over a thousand geothermal sites in the U.S. that have not yet been developed, but many are remote from population centers and potential grid connections. Geothermally produced electricity can be designed provide continuous, baseload power though this is usually not the case. A study by the Massachusetts Institute of Technology suggested that roughly 10% of U.S. electrical energy could be supplied by harnessing accessible geothermal resources at costs that are competitive with fossil fuel generation.

Waste-to-Energy

There has been an ongoing debate among energy management professionals as to whether or not waste-to-energy (WTE) systems can be generally classified as renewable energy systems. If the source of the waste for the WTE system is woody biomass from a tree-harvesting process, then the consistency of the waste and its source qualify it as a waste from biomass, a renewable energy source. If the energy is sourced from municipal mixed wastes, then it can be considered to be partially renewable. According to the UK's Department for Environment Food and Rural Affairs, "energy from residual waste is only partially renewable due to the presence of fossil-based carbon in the waste, and only the energy contribution from the biogenic portion is counted towards renewable energy targets" [22].

Energy derived from wastes is considered to be a more sustainable alternative than placing it in landfills, providing the residual waste being used has the right renewable content and is matched with a process that is efficient at turning the waste into energy [22]. It also produces a lower cost, domestic source of heat which can be used to produce electricity for microgrids. For municipal mixed wastes, these partially renewable energy sources can provide heat, electricity, and transportation fuels.

WTE plants are used for combined heat and power or electrical generation applications. A WTE plant owned by Pinellas County, Florida operates as an enterprise fund by charging capacity payments, electricity sales, material sales, and tipping fees [23]. The plant generates electricity and reduces the amount of waste that must be place in a sanitary landfill. It provides a heuristic example of how these plants are configured and how the business model supporting them can be successful. Approximately 80% to 90% of all municipal solid waste from the county is directed through the Pinellas County Solid Waste Integrated Solid Waste Management Facility [23]. This solid waste stream provides a constant flow of fuel for the plant and allows it to operate without using taxpayer funds [23]. The plant processes garbage for the combustion process and uses the heat to boil water to make steam, which then turns a turbine to make electricity (see Figure 6.7) [23]. The key benefit to burning trash is that its volume is reduced by nearly 90% and its weight by 70% [23]. The plant is designed with three boilers that operate 24 hours per day producing high-pressure steam (610 psig and 750°F) to rotate two turbines that are connected to two generators with rated capacities of 25 MW and 50 MW [23]. Combined, the two generators can produce steam at the rate of 700,000 lb./hour. The plant burns roughly 3,000 tons of garbage every day, or almost one million tons annually [24]. Natural gas burners (two per boiler) supplement the combustion process to reduce carbon monoxide emissions from the waste fuel source [23]. Emissions from the incinerator are treated and filtered before they reach the stack and are emitted into the atmosphere [23]. About 60 MW of electricity is sold to Duke Energy, enough to power about 40,000 homes and businesses [24].

COMPARITIVE COST OF GENERATION SYSTEMS

Renewable generation for microgrids provides opportunities to lower the cost for electricity when compared to fossil fuels. However, they can operate more efficiently and reduce costs to a greater extent when the microgrid smooths the variation in generation [26]. For example, the initial capital costs for the construction of geothermal power plants are higher than for new coal-fired or natural gas turbine plants [21].

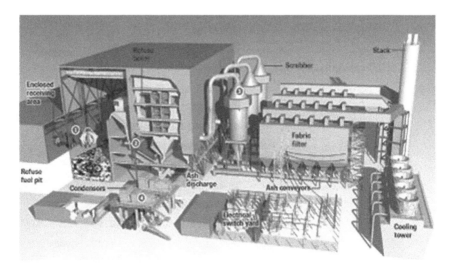

FIGURE 6.7 Schematic diagram of the Pinellas County Solid Waste Treatment Facility [25].

However, geothermal plants have reasonable operation and maintenance costs and no fuel costs, substantially lowering their lifecycle costs [21]. Geothermal plants can be used to supplement other renewable resources, such as solar or wind power that provide intermittent generation, by supplementing their operations with baseload capacity.

When comparing plant costs, overnight capital cost is often used as a basis for comparisons in the power industry. This refers to the cost of building a power plant as if it were to somehow happen overnight. The term is useful when comparing the economic feasibility of building various types of generating plants and is expressed in $/KW per unit of capacity. As Figure 6.8 indicates, distributed generation technologies are competitive both for base and peak loading generation. Renewable technologies including solar, wind, hydropower, and geothermal compare favorably to coal with sequestration and nuclear power.

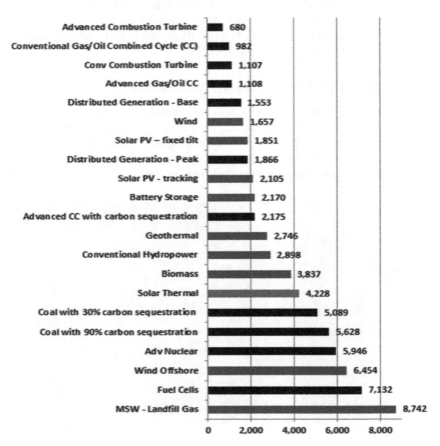

FIGURE 6.8 Estimated capital cost of generating technologies. (Source: Energy Information Administration [27].)

SUMMARY

Providers of renewable energy sources such as solar PV and wind power have scaled their technologies, improved their capacities for generation, and driven down installation costs [14]. Meanwhile, prices for electricity from renewable energy systems have fallen rapidly, making them more affordable and realistic as options for distributed generation [14]. This broadens their commercial appeal especially for microgrids. The use of renewable energy is both a solution for microgrid generation applications and a driver for microgrid development. Microgrids can be powered using hydropower, solar thermal or PV, landfill gas, biomass, wind power, and waste-to-energy systems.

There is enormous potential when microgrids are designed to incorporate renewable energy generation. As added generation capacities have leveled off in developed countries, they are increasing in the world's developing countries where renewables have become the primary source of new electrical generation. In 2017, the majority of all power generation capacity (totaling 186 GW) was from renewables (totaling 94 GW) [28]. According to the International Energy Agency's market forecast report, the world's total renewable energy capacity is expected to increase by 50% between 2019 and 2024 adding 1,200 GW—equivalent to the current total power capacity of the U.S.—with solar PV accounting for 60% of the increase [29]. The share of renewables in global power generation is predicted to rise from 26% today to 30% in 2024 [29]. Increased investments in developing countries is making renewables more competitive with fossil generation by lowering the costs of renewables, especially wind power and solar generation [28].

Due to technological improvements and falling costs, wind and solar generation are becoming comparable in cost with conventional generation sources in a number of markets [30]. They are seen as the primary solution in parts of the world that lack transmission infrastructure. As a result, renewable generation is becoming preferred for microgrid applications in previously underserved off-grid locations. However, despite the growth of market-driven capacity additions, some of the recent renewable electricity growth has been due to the availability of financial incentives such as tax credits and feed-in tariffs [30].

The IEA estimates that 33 million people have electricity access with off-grid renewables and mini-grid solutions [30]. The pace of solar electrification has been accelerating with an estimated five million people having gained access each year since 2012, compared with about one million on average between 2000 and 2012 [30]. Regardless, North America remains the leader in microgrid development. The future potential for technology transfer and new markets for renewable energy-based microgrids is increasing.

The U.S. has a diverse, abundant, and geographically distributed set of renewable energy resources. The electrical power sector is being transformed with the incremental additions of generation from renewables, especially geothermal energy, wind power, and solar energy. As renewable generation technologies are more widely understood and obtain regional cost-parity with fossil fuel generation, this transformation to clean energy will accelerate the opening of new markets for microgrids.

Renewable energy microgrids are versatile and are being encouraged by some regional electric utilities as they can be deployed rapidly for special situations. Southern California Edison constructed a solar microgrid in 27 days in cold and windy weather to solve an urgent need to avert and inspect wildfires near a remote dam in California's Sierra Nevada mountains [31]. The microgrid, completed in 2019, powered valves, emergency backup, cameras, and other equipment needed to ensure that the Gen Dam (part of the Rush Creek Hydroelectric Facility near Yosemite National Park) was operating properly; it included two 3.3 kW solar arrays, a 600-amp, 48-volt storage system, and a propane backup generator [31].

The *Renewable Electricity Futures Study*, by the National Renewable Energy Laboratory (NREL), indicated that if renewable generation were combined with a more flexible electrical grid, they could provide 80% of the needed U.S. electricity supply by 2050 [32]. The study further noted that "greater flexibility will enable operators to more readily maintain the required balance between electricity supply and demand, even at high levels of variable renewable generation on the grid" [32]. For microgrids, flexibility can be provided by an energy generation portfolio that includes hybrid renewable energy systems and improved microgrid controls. As the costs for the renewable energy systems used to generate electricity decline, the costs of coal-fired and nuclear plants continue to increase giving developers more reason to consider microgrids. Simulations and analysis of an integrated network of generation, transmission, and storage have shown that they are capable of reliably meeting the need for electricity across the entire U.S. [32]. Renewables when ideally configured have the surprising potential capability to meet electricity demand for every region of the U.S. every hour of every day [32].

REFERENCES

1. NextGen Climate. The economic case for clean energy, page 12. https://nextgenamerica.org/wp-content/uploads/2015/07/cleanenergyreport.pdf, accessed 18 May 2020.
2. Energy Information Administration (2018, November). Electric power monthly with data for September 2018.
3. Hill, J. (2018, November 30). Renewables account for 18% of U.S. generation. https://cleantechnica.com/2018/11/30/renewables-account-for-18-of-us-electricity-generation, accessed 2 January 2020.
4. Dwivedi, R. (2016, August 23). Smart grid-innovation ahead. https://ieeeuiettechscript.weebly.com, accessed 2 December 2019.
5. Jiankun, H. (2018, November 26). A holistic response – China to promote global collaboration at UN climate change event in Poland. *Time*.
6. Roosa, S., ed. (2018). *Energy Management Handbook*, 9th edition. Fairmont Press: Lilburn, Georgia, pages 453–460.
7. Purdue University, State Energy Forecasting Group, Energy Center at Purdue University (2018, October). 2018 Indiana renewable energy resources study. https://www.purdue.edu/discoverypark/sufg/docs/publications/2018_RenewablesReport.pdf, accessed 18 May 2020. West Lafayette, Indiana.
8. Davidson, P. (2008, October 28). Water power gets new spark. *USA Today*, page 33.
9. Bruggers, J. (2008, October 28). Retrofit hydro plant offers clean power. *The Courier-Journal*, page E1.

10. Power Technology (2010, January 19). Ethanol power plant, Minas Gerais. https://www.power-technology.com/projects/ethanol-power-plant, accessed 2 January 2020.
11. Quizlet. Solid and hazardous waste. https://quizlet.com/75747936/solid-and-hazardous-waste-flash-cards, accessed 16 December 2019.
12. U.S. Environmental Protection Agency (2017, February 14). Understanding global warming potentials. https://www.epa.gov/ghgemissions/understanding-global-warming-potentials, accessed 16 December 2019.
13. U.S. Environmental Protection Agency (2019, July 3). Benefits of landfill gas energy projects. https://www.epa.gov/lmop/benefits-landfill-gas-energy-projects, accessed 16 December 2019.
14. U.S. Environmental Protection Agency (2019, July 30). Basic information about landfill gas. https://www.epa.gov/lmop/basic-information-about-landfill-gas, accessed 12 December 2019.
15. Renewable Energy Access (2007, June 27). 700 MW of electricity to come from landfill gas. https://www.renewableenergyworld.com/2007/06/27/700-mw-of-electricity-to-come-from-landfill-gas-49123, accessed 7 January 2020.
16. Mey, A. (2018, September 25). Electricity monthly update, U.S. solar capacity consists of many small plants. U.S. Energy Information Administration. https://www.eia.gov/electricity/monthly/update/archive/september2018, accessed 1 January 2020.
17. Marcy, C. (2018, October 26). Today in energy. U.S. Energy Information Administration. https://www.eia.gov/todayinenergy/detail.php?id=37372, accessed 1 January 2020.
18. Gupta, A. (2017, March 18). First Solar begins operation of 250 Megawatt Moapa Southern Paiute Solar Project. https://www.eqmagpro.com/first-solar-begins-operation-of-250-megawatt-moapa-southern-paiute-solar-project, accessed 7 January 2020.
19. U.S. Department of Energy (2010, October). Energy efficiency and renewable energy. DOE/EE-0353.
20. Patel, S. (2019, January 16). GE, Siemens separately announce developments for mammoth offshore wind turbines. *Power Magazine*. https://www.powermag.com/ge-siemens-separately-announce-developments-for-mammoth-offshore-wind-turbine/?mypower, accessed 17 January 2019.
21. Pace Energy and Climate Center (2000). Power scorecard: electricity from geothermal energy. http://www.powerscorecard.org/tech_detail.cfm?resource_id=3, accessed 27 January 2019.
22. Department for Environment Food and Rural Affairs, UK (2014, February). Energy from waste, a guild to the debate. https://assets.publishing.service.gov.uk/government/uploads/system/uploads/attachment_data/file/284612/pb14130-energy-waste-201402.pdf, accessed 22 November 2019.
23. Guimond, M. (2019). Pinellas County waste-to-energy plant case study. *International Journal of Energy Management*, 1(1), pages 66–75.
24. Pinellas County. Waste-to-energy and industrial water treatment facilities. https://www.pinellascounty.org/solidwaste/wte.htm, accessed 23 November 2019.
25. Oswald, K. (2015, December 17). Management of waste in Pinellas County. www.stpete.org/sustainability/docs/3A%20and%205B_WTE_%20Resource%20Management%20in%20Pinellas%20121715.pdf, accessed 18 November 2018.
26. S&C Electric Company (2018). Is a microgrid right for you? Part 1 of 3. https://www.sandc.com/globalassets/sac-electric/documents/sharepoint/documents---all-documents/education-material-180-4504.pdf, accessed 1 January 2020.
27. U.S. Energy Information Administration (2018, February). Cost and performance characteristics of new generating technologies. *Annual Energy Outlook 2018*.
28. BloombergNEF (2018, November 27). Emerging markets outlook 2018. http://global-climatescope.org/assets/data/reports/climatescope-2018-report-en.pdf, accessed 18 May 2020.

29. International Energy Agency (2019, October 21). Global solar PV market set for spectacular growth over next 5 years. https://www.iea.org/newsroom/news/2019/october/global-solar-pv-market-set-for-spectacular-growth-over-next-5-years.html, accessed 2 December 2019.
30. International Bank for Reconstruction and Development (2018). Tracking SDG7: the energy progress reports 2018. https://trackingsdg7.esmap.org/data/files/download-documents/key_messages.pdf, accessed 19 January 2018, pages 29–74.
31. Cohn, L. (2019, December 13). California utility builds solar microgrid in 27 days to avert wildfires in Sierra Nevada. https://microgridknowledge.com/solar-microgrid-sce-wildfires, accessed 24 January 2020.

7 Energy Storage Technologies for Microgrids

ENERGY STORAGE SOLVES PROBLEMS

Energy storage systems enhance the capacities of microgrid systems. They are fast becoming an integral component of renewable generation systems to resolve issues with intermittency. Some definitions of *microgrid* specifically note electrical storage as an integral component of microgrids. For example, the CIGRE C6.22 working group defines microgrids as "electricity distribution systems containing loads and distributed energy resources (such as distributed generators, storage devices, or controllable loads) that can be operated in a controlled, coordinated way either while connected to the main power network or while islanded" [1].

The often-cited problem with the generation of electricity is that it must be used the instant it is generated. If the energy cannot be used at the instant of production, it cannot perform work and has no economic value. Another problem is that electricity cannot be directly stored. When not consumed instantaneously, to be useful it must be converted to another form of energy which can be stored for later use. Energy losses and inefficiencies result from the processes that convert the electricity into energy that is storable and reconvert the energy back into useful electricity. The percentage of energy that is useful at the end of the process is sometimes referred to as the storage system's *turn-around efficiency* or *round-trip efficiency*.

When generation systems lack energy storage and supply excess electricity directly to the grid, these losses are avoided but distribution system losses may be increased. Electrical energy storage systems (ESS) can serve either on-site, behind-the-meter, or utility grid applications. Most non-utility microgrid storage solutions are behind-the-meter applications. Energy storage to support microgrid functions may be required for both short-term and long-term requirements.

Electricity storage can be deployed at or near any of the major subsystems in an electric power system: generation, transmission, substations, distribution, and at consumer sites [2]. Systems that store electrical energy serve both as storage (during the charging process) and generation (when stored energy is regenerated and released in the form of electricity). For this reason, electrical energy storage systems are often functionally classified as being electrical generation systems. Traditional uses for electrical energy storage include: 1) improving the reliability of the power supply since storage systems support users when power network failures occur due to natural disasters; 2) maintaining and improving power quality, frequency, and voltage; and 3) reducing electricity costs by storing electricity obtained during off-peak times when its price is lower and discharging electricity during peak periods when its price is higher instead of purchasing higher cost electricity [3].

To be successfully deployed, microgrids must be capable of providing a fully integrated electrical production system which is often enhanced by a utility interconnection. Energy storage systems provide a number of benefits when incorporated into electrical grids; they: 1) enable time-shift of energy delivery; 2) supply capacity credit; 3) provide grid operational support; 4) provide transmission and distribution support; 5) help maintain power quality and reliability; and 6) allow improved integration of intermittent generation [2].

For microgrids, the benefits of energy storage systems include grid benefits plus enable digital controls and smaller-scale energy storage systems to provide more consistent voltage and frequency with more reliable kVAR control [4]. Energy storage is considered to be among the most promising technologies for mitigating voltage and/or frequency deviations [5]. When a voltage dip occurs, the charged energy storage system can rapidly feed energy back into the microgrid to provide stability [4]. Energy storage also supports electrical generators by accepting block loads without fluctuations in frequency [4]. This capability makes it possible to deploy renewable resources in far greater proportions than a utility grid could support—up to and exceeding total system demand [4]. Some energy storage systems can supply power more rapidly and at lower cost than would be required to bring additional non-intermittent generation online.

Like central utility grids, microgrids must resolve issues associated with intermittent electrical loads. While central grids are likely to balance supply and demand for power with excess capacity, microgrids are more likely to incorporate energy storage and load response systems. For microgrids, the presence of intermittent loads and variable sources with power-electronic converter interfaces may lead to power quality issues. These issues are most often voltage variations, frequency variations, and waveform distortions which are commonly referred to as harmonic distortions [5]. Passive filtering can be used to reduce harmonic distortions, which may require capacitors and inductors. Active filtering is emerging as an alternative that could be realized within the interfacing converter of an energy storage system [5].

ELECTRICAL ENERGY STORAGE

One common problem with electrical generation has to do with having adequate supplies available to meet periods when electrical demand loads are highest. To satisfy the peak load requirements, additional generation must be available or loads must be reduced. Maintaining the additional generation capacity online to meet peaking needs requires additional energy consumption. When the electricity is generated using fossil fuels, atmospheric carbon emissions result. When demand moderates, generators must still remain online, thus producing more power than is needed. Consider all the greenhouse gas emissions that could be reduced if additional generating capacity (new power plants) were not required for peak periods. There are ways to store large amounts of electrical power and to release it during peak load conditions. Energy storage systems provide load-leveling capabilities, increase microgrid capacities, and help ensure uninterrupted service.

Energy Storage Technologies for Microgrids

There are many uses for storage systems. They can provide energy balancing by shifting and leveling loads, help meet peak demand situations by releasing spinning reserve power, and increase capacities. For example, standalone microgrids that have solar photovoltaic (PV) systems often generate excess electricity during daylight hours. Since the demand for the power is less than the output of the system, battery banks can be used to store the power until needed, improving system capacity. The batteries can serve multiple purposes such as stabilizing current and voltage, and providing surge currents to variable loads when needed. Examples of variable loads include electric motors, pumps, and fans.

The applications of storage technologies for microgrids are similar to those of the larger utility companies. Long-duration and short-duration applications require capacities for prolonged discharges and include time-shift of energy delivery, operational support, power quality, reliability, and integration with intermittent sources of energy [2]. Examples of long-duration applications include mitigating transmission curtailments, time-shifting renewable generation, and load shifting. Short duration applications include regulation control, fluctuation suppression, frequency excursion suppression, and grid voltage stability [2]. Other storage applications that can be mitigated are infrequent or frequent discharges. Energy storage can be accomplished by installing storage systems at the utility substations, community facilities, residences, and commercial end-user facilities (see Figure 7.1).

There are a number of different types of battery storage systems that are used for microgrids. These include mechanical energy storage, electrical/electrochemical energy storage, chemical energy storage, biological energy storage, and thermal energy storage.

MECHANICAL ENERGY STORAGE

Common mechanical energy storage systems include pumped storage hydroelectricity, compressed air energy storage (CAES), and flywheel energy storage.

PUMPED STORAGE HYDROELECTRICITY

Hydroelectric systems typically have a dam that stores water in a single reservoir. When electricity is needed, gates open that allow the water to flow to the turbines. Water exits via the tailraces and enters the river below the dam. To reuse the same water to generate additional hydroelectric power, the primary strategy is to construct additional dams downstream to form more reservoirs at lower elevations.

Pumped hydro storage is an alternative that recycles the water repeatedly at the same site. A pumped storage hydroelectric plant uses two or more reservoirs, with at least one located at a much higher elevation than the other. The reservoir at lower elevation is called the *lower reservoir* and the one at higher elevation is called the *upper reservoir*. The storage concept involves reusing the water from the lower reservoir (after it has passed through the turbines) by pumping it to the upper reservoir to generate additional electricity. This type of energy storage system uses excess electricity, usually generated by the hydropower site, to pump water to higher altitude where it is stored as gravitational potential energy [2]. During periods of low

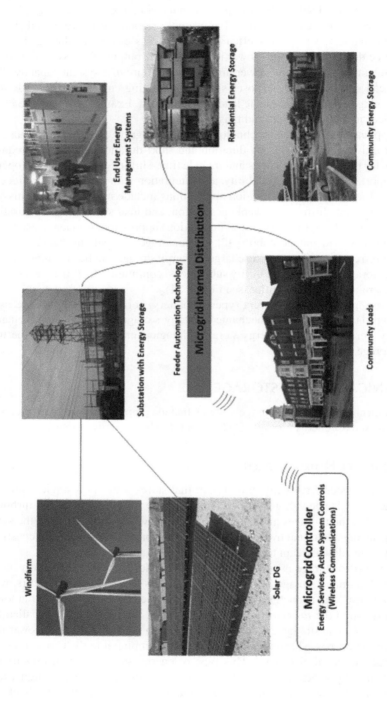

FIGURE 7.1 Microgrid configuration with both substation and community energy storage.

demand for electricity, such as nights and weekends, water is stored by reversing the turbines (or providing an alternative water pumping system) and pumping the water from the lower to the upper reservoir. The stored water is later released, passes through the penstock, turns the turbines, and generates electricity as gravity causes it to flow back into the lower reservoir. The round-trip efficiency of pumped hydro systems is between 75% and 78% [2]. The energy losses are mostly due to the inefficiency of pumping water from a lower to a higher elevation against the force of gravity. However, pumped storage plants have higher operating costs than conventional hydropower plants due to the added equipment and costs associated with pumping water uphill to refill the upper reservoirs [6].

Pumped storage hydropower systems are generally either open-loop or closed-loop systems. An *open-loop system* has a continuous source of downstream water that is pumped uphill to an upper storage reservoir, typically pumping water that has already passed through the dam up to the storage reservoir above the dam [6]. *Closed-loop systems* pump water from a lower storage reservoir that is not continuously filled with water and is not connected to a flowing source of water [6].

To make the investment of constructing two reservoirs for pumped hydro storage systems economically feasible, key variables include plant costs, power availability, and the value of electricity at different time periods. These systems have the greatest economic efficiency when there is a wide differential in the value of electricity at low demand compared to high demand coupled with an abundance of power available at low demand. While the cost difference to make pumped hydroelectric storage economically attractive varies, price multiples for electricity for peak demand periods compared to off-peak of five to ten times are considered reasonable.

Interestingly, of the range of electric storage system technologies available, the majority of utility-supported storage systems use pumped hydroelectric storage with a total of 176 GW installed in 2017. In the U.S., most pumped storage systems were constructed between 1960 to 1990 (see Figure 7.2) [6]. California has the most pumped storage capacity, with 3.9 GW, or 17% of the national total; other states including Virginia, South Carolina, and Michigan have at least 2 GW of hydroelectric pumped storage capacity [6]. In terms of total installed capacity, pumped hydro is the predominate technology used to provide energy storage for microgrids.

The Gorona del Viento (GdV) El Hierro project, in the Canary Islands, is one of the world's largest standalone hybrid, fully integrated, renewable energy power generation systems. The project's goals were to: 1) reduce dependency on diesel-fired generation by supplying all of the power required for the island's 11,000 residents (about 38.7 MWh annually); and 2) scale the microgrid to provide greater operational stability and security. It incorporates wind power (11.5 MW) and solar PV generation with pumped hydro storage (200,000 m^3 of storage or 11.3 MW) which stores surplus wind energy by pumping water up 700 meters (2,300 feet) to a lake located in the crater of an extinct volcano. This island microgrid project started production in June 2015 and has since supplied the island with almost half of its total electricity requirements [7]. Table 7.1 provides a summary of the project's electrical generation performance from January 2016 through December 2017, the first two years of operation. The renewable output increased in 2017 from 2016 as operators fine-tuned

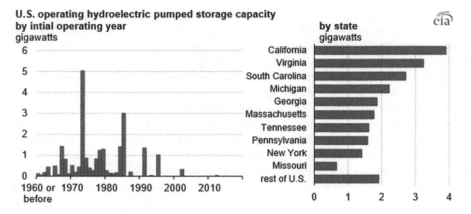

FIGURE 7.2 U.S. pumped storage installations. (Source: U.S. Energy Information Administration [6].)

the controls and multiple generation systems involved. The share of electricity generation from renewables is greater during the summer months as output from the solar plant increases. In the third quarter of 2018, the GdV project successfully supplied 100% of El Hierro's electricity for a total of 46 days, with 18 days of continuous 100% generation during a sustained period of high winds [7].

Compressed Air Energy Storage

Compressed air storage (CAES) has proved to be an affordable technology used to provide power and pressure for mechanical systems. Storage tanks with sufficient volume are an important component of compressed air systems. A storage tank is used as an air reservoir that allows system pressure to remain relatively steady even during air-demand surges and swings [8]. Without proper storage, the compressed air system may experience sudden pressure losses during any sharp air-demand increases throughout the day [8]. Installing storage tanks is a more energy-efficient strategy to satisfy momentary demand surges; plus they prevent compressors controlled by an on/off or load/unload methods from rapidly cycling [8]. Longer cycles allow a load/unload compressor's sump more time to release pressure and operate at a lower sump pressure while unloaded, thus saving energy [8]. Increased storage allows compressors in inlet-modulation or variable-speed modes to maintain a steadier output profile [8]. Large-scale CAES applications must be designed to conserve the heat energy associated with compressing air since dissipating heat lowers the energy efficiency of the storage system [9].

CAES plants can be used for microgrid storage applications. However, variable loads can cause voltage and frequency instabilities [10]. To maintain system stability, microgrids must generate power based on present conditions [10]. Compressed air plants are similar to pumped hydro powerplants in terms of their applications, output, and storage capacities [11]. Instead of pumping water between reservoirs during periods of excess power, a CAES plant compresses ambient air and stores

TABLE 7.1
Electrical Generation of the Gorona del Viento El Hierro Microgrid Project for 2016–17

Year/Month	Renewable generation (wind & solar) MWh	Renewable generation (wind & solar) %of total	Fossil fuel generation (diesel) MWh	Fossil fuel generation (diesel) %of total	Total generation MWh
2016					
January	849	22.2	2,971	77.8	3,820
February	1,860	53.5	1,615	46.5	3,475
March	1,572	40.2	2,343	59.8	3,915
April	1,325	36.3	2,330	63.7	3,655
May	962	25.4	2,327	74.6	3,789
June	2,099	53.9	1,793	46.1	3,892
July	2,635	66.0	1,336	34.0	4,071
August	2,340	53.4	2,038	46.6	4,378
September	2,266	56.6	1,739	43.4	4,005
October	767	19.8	3,111	80.2	3,878
November	1,060	29.3	2,559	70.7	3,619
December	1,076	27.1	2,394	72.9	3,970
Total	18,861	40.6	27,606	59.4	46,467
2017					
January	1,146	29.9	2,691	70.1	3,837
February	1,518	44.5	1,891	55.5	3,409
March	2,231	57.5	1,648	42.5	3,879
April	1,041	27.5	2,745	72.5	3,786
May	1,555	40.8	2,254	59.2	3,809
Jure	2,315	61.2	1,467	38.8	3,782
July	3,234	78.3	896	21.7	4,130
August	2,562	57.8	1,372	42.2	4,434
September	2,436	62.4	1,465	37.6	3,901
October	1,039	26.1	2,947	73.9	3,986
November	853	24.7	2,594	75.3	3,447
December	1,362	37.6	2,262	62.4	3,624
Total	21,32	46.3	24,732	53.7	46,024

Source: data from [7].

it under pressure in tanks, underground caverns, or depleted natural gas reservoirs [11]. This enables a CAES to be used to balance fluctuations in power generation and consumption since the excess power compresses and stores air [10]. When electricity is required, the pressurized air is heated and expanded in an expansion turbine driving a generator for power production [11]. The CAES uses an air flow controller to enable the microgrid to track the various load demands and maintain a stable frequency [10].

Germany's Huntorf compressed air energy storage plant is the world's first and still largest utility-scale CAES plant. The 321 MW plant has operated since 1978, functioning primarily for cyclic duty and ramping duty, and as a hot spinning reserve for the industrial customers in northwest Germany. The plant levels the variable power from numerous wind turbine generators in Germany. This plant and one in McIntosh, Alabama both use a diabatic CAES method, where the compression of the combustion air is separated and independent from the actual gas turbine process [11].

Flywheel Energy Storage

Flywheel energy storage systems (FESS) store and release energy due to torque. FESS use a flywheel which typically spins at a very high rate in a vacuum and stores energy as rotational energy [12]. This spinning mass, called a rotor, rests on bearings that facilitate its rotation [2].

The components of these systems include a rotor and bearings that are housed in a sealed containment that reduces friction between the rotor and the surrounding environment and provides a safe guard against hazardous failure [2]. A typical FESS flywheel has a shaft with a rotor attached and a motor generator that both drives the shaft and extracts energy [13]. Modern flywheels use rotors made of composite materials and are capable of spinning at speeds of 100,000 rpm or higher [13]. Some have massive metal rotors but the rotational speed of these is limited by their ability to resist higher centrifugal forces [13]. Flywheels use either mechanical bearings or magnetic bearings (or super-conducting magnets) to reduce friction and improve durability [12]. Increased storage efficiency is achieved by reducing friction when a flywheel is used to store energy for longer periods of time since mechanical flywheels dissipate most of their energy during the first hour after charging [12]. To minimize energy losses, most newer flywheels use magnetic bearings and operate in a vacuum chamber [13]. For charging, electricity powers a motor-generator that spins a shaft connected to the rotor to store energy [2]. When discharging, the stored kinetic energy is converted into electricity by allowing the momentum to power the motor-generator [2].

Flywheel energy storage systems can be used in microgrid applications to provide a boost in power output during periods when peaking power is needed and more costly that baseload power.

ELECTRICAL/ELECTROCHEMICAL ENERGY STORAGE

Types of electrical/electrochemical energy storage systems include capacitors, batteries, fuel cells, and grid storage. Battery systems are sometimes used by utilities as an operating reserve. When a sudden drop of electric supply happens, battery storage systems provide reserve power until large-capacity generation becomes available [14]. Batteries can be either vented cell (wet cell) or sealed cell (gel cell) [15]. In sealed cell batteries there is a minimum amount of electrolyte absorbed in a gel; they require less maintenance than traditional wet cell batteries as there is no need to add water [15]. Batteries used for stationary power systems (SPS) must perform well at low and high temperatures, and have a high energy density and long cycle life for deep discharge applications [15].

Battery storage systems can be used in microgrids for ancillary services, load shifting, renewable energy integration, and behind-the-meter applications. The types of services that battery systems offer include providing frequency response, reserve capacity, black-start capability, power storage for electric vehicles, mini-grid upgrades and support for self-consumption of rooftop solar power [16].

Lead-Acid Batteries

The lead-acid batteries used for microgrids are similar to common automobile batteries used in vehicles with internal combustion engines. They are among the most ubiquitous batteries used today and almost all components can be recycled. Lead-acid batteries are commonly used to provide electrical storage for reserve power in remote applications such as telecommunications sites. However, for utility applications they require a warehouse-full of interconnected batteries to store electricity.

During charging cycles, lead-acid batteries generate heat which must be dissipated. To provide the best reliability, reserve batteries need to be selected for high temperature applications [17]. When rated for such conditions, the batteries have a theoretical maximum ten-year life when operated at 25°C (77°F) [17]. In practice, the lifecycle of lead-acid batteries is much less (about five years) due to widely variable ambient conditions. Typically, a lead-acid battery loses 50% of its service life for each 10°C increase in ambient temperature above its normal temperature rating [17]. While relatively inexpensive, lead-acid batteries have a low energy density, are heavy, often do not respond well to deep discharging, and since the lead (Pb) is toxic their use may be restricted in some applications or locations [16]. The Pb in these batteries contaminates the environment if disposal is not properly handled. For these reasons, lead-acid batteries are being replaced with lithium-ion batteries for many applications.

Lithium-Ion Batteries

Lithium-ion (LI) batteries exchange lithium ions (Li+) between the anode and cathode, both of which are made using lithium intercalation compounds [16]. For example, lithium cobalt oxide ($LiCoO_2$), originally introduced in the 1980s, was the active positive material in the original LI battery designs [16]. Advantages of LI batteries include higher specific energy, lighter weights, higher energy, and more power density compared to other battery technologies, ability to provide high-power discharge capability, excellent round-trip efficiency, a relatively long lifetime, and a low self-discharge rate [16]. The lifecycle of LI batteries varies widely depending on cell design and operating conditions, ranging from 500 to 20,000 full cycles depending on the physical design and type of technology used [16]. Lithium-ion batteries offer good charging performance at cooler temperatures and provide fast-charging within a temperature range of 5°C to 45 °C (41°F to 113°F) [18]. Prismatic LI batteries are among the largest types. They are used for electric vehicles and in applications previously supported by lead-acid batteries including backup power and off-grid telecom systems [19]. For renewable energy storage, large-format prismatic lithium iron phosphate (LiFePO$_4$) batteries are often used [19].

The costs for LI battery systems have been declining. In 2015, LI batteries ranged from $1,000 to $2,000 (U.S.) per kW. By late 2019 prices for vehicular LI systems had dropped below $200/kW. While considered to be an expensive storage option, it is anticipated that LI batteries for utility applications will decline from about $1,200 per kWh today to about $400 per kWh by 2035 [20]. This trend in declining costs has been noticed by electric utilities. In Arizona, the 20 MW Pinal Central Solar Energy Center coupled a utility-scale 20 MW solar plant (258,000 panels with tracking capability) with 10 MW of LI battery storage systems, a project that powers the equivalent of 5,000 homes. The Hornsdale Power Reverse, installed in 2017 in South Australia, is the world's largest LI battery storage facility with a 100MW/129MWh capacity.

SODIUM SULFUR BATTERIES

Sodium and Sulfur (NaS) battery banks can store large amounts of electricity. High-temperature batteries such as NaS batteries use liquid active materials and a solid ceramic electrolyte made of beta-aluminum that also serves as the separator between the battery electrodes [3]. Typically, the anode material in these systems is molten sodium and the anodes rely on sodium-ion transport across the membrane to store and release energy. The cathode for the most common NaS battery configuration is molten sulfur [3].

The development of NaS battery banks eliminates the inconveniences and obstacles associated with lead-acid batteries. By bridging electrodes with a porcelain-like material, a room-sized bank of durable NaS batteries will last approximately 15 years. Use of NaS battery banks also reduces the threat of environmental impacts. For future microgrid applications, NaS batteries hold potential to transform how electricity is stored and delivered to meet peak power demands [21].

According to Stow Walker of Cambridge Energy Research Associates, by "using NaS batteries, utilities could defer for years, and perhaps avoid, construction of new transmission lines, substations and power plants" [21]. This is possible since the NaS batteries can be charged during periods of low electrical demand, such as nights or weekends when electrical costs are lowest, and then discharge power during peak demand conditions. The battery banks can be more conveniently located near existing substations. Electric utilities benefit since the battery banks can provide backup power during power outages and be used in combination with solar and wind power production [21]. For microgrids, their modularity enables them to be scaled down and sized to meet variable load conditions. In these applications, electricity stored during peak production periods can be released on demand when additional capacity is needed.

Costs for NaS batteries have dropped considerably. In 2007, the cost was approximately $2,500 (U.S.) per kilowatt or about 10% more than the cost of building a new coal-fired power plant to produce electricity [21]. Today, the costs for the total system installation for an NaS battery energy storage system ranges between $263/kWh and $735/kWh [16]. This rapid cost decline has opened new markets for NaS systems such as those for microgrid applications. A disadvantage of NaS storage systems is their relatively high annual operating cost which can be $40 to $80/kW [16]. However, their advantages far outweigh the added costs. NaS batteries are much smaller in size than most other batteries, have high round-trip efficiencies, and all components can be recycled. They do not require additional fuel and do not emit greenhouse gasses [21].

NaS battery systems are a proven technology that is being used by electric utilities in the U.S. and Japan. Completed in 2008, the Rokkasho-Futamata Wind Farm was the largest and among the first combined wind generation (51 MW) plus battery energy storage (34 MW) facility to be completed in Japan. It houses one of the world's largest sodium sulfur battery assemblies and uses the NaS batteries for load leveling, enabling the storage of low-cost, off-peak wind power for sale or distribution during peak demand times. The energy storage cycle efficiency of the system is 89% to 92% which is higher than pumped hydro.

VANADIUM REDOX BATTERIES

Vanadium redox batteries (VRBs) are flow batteries that were pioneered in the Australian University of New South Wales (UNSW) in early 1980s. The Australian Pinnacle VRB bought the basic patents in 1998 and licensed them to Sumitomo Electric Industries (SEI) and VRB Power Systems. VRBs have long lifecycles, high energy efficiencies (up to 85%), long durations (1–20 hours) with continuous discharge and high discharge rates, quick response times, and heat extraction capabilities [16]. Low electrolyte stability and high costs are among the disadvantages [16].

VRB systems with storage capacities up to 500 kW (5MWh) have been installed in Japan. VRBs have also been used in power quality applications. Japan's Tomamae Wind Villa Power Plant (34.6 MW), completed in 2005 at a cost of $72 million, continues to be one of the world's largest vanadium redox flow battery energy storage installations. At the time of commissioning it was Japan's largest wind power plant. The vanadium flow battery system was installed at the existing Tomamae Wind Villa with the intent to provide wind power smoothing for the intermittent and variable wind plant.

ZINC-AIR BATTERIES

Metal-air energy storage systems offer a long life in a safe and non-toxic package. A rechargeable capability has been developed for the zinc-air energy storage technology common in applications such as backup power for cellular communication towers [4]. The batteries can be 95% discharged and can be recharged with no cycle limit. Rechargeable zinc-air battery systems are among the most economical electricity storage technologies and include integrated controls with monitoring at the cell level [4]. Zinc-air batteries do not overheat or discharge dangerous concentrations of hazardous gases and they operate in a range from 0°C (32°F) to 50°C (122°F) without derating [4]. Life expectancy is at least twice that of lead-acid batteries. The next generation of zinc-air storage will be offered in capacities at the megawatt scale, well suited for hybrid microgrids, and will provide an attractive total cost of ownership [4].

HYDROGEN ENERGY STORAGE

Common forms of chemical energy storage systems include hydrogen (H_2), biofuels, and fossil fuels. Hydrogen energy storage, used in vehicular transportation systems, is rarely found in microgrid systems. Today, fossil fuels (typically natural gas, oil, and coal) are primarily used in both utility-scale electrical systems and microgrid applications to generate the needed H_2. This may change in the future as renewable hydrogen technologies become more commonplace.

Hydrogen gas offers promise as a form of energy storage if derived in a sustainable manner, meaning from some way other that using a fossil fuel. One methodology for making renewable hydrogen is to produce the H_2 from a biofuel. Another is using solar–hydrogen production technologies. The process uses solar-generated electricity to split water into its components (H_2 and O_2) via electrolysis, then stores the extracted hydrogen for later use. The hydrogen can then be transferred from storage and used to produce electricity using fuel cells or by an engine that converts the energy into electricity. Though the use of this technology is not widespread, there are microgrids that use H_2 electrolysis processes in about 30 countries. Hydrogen can also be used in vehicles as the combustion process is similar to using gasoline. A benefit of using renewable hydrogen as a fuel is that it produces only negligible pollution.

One strategy for microgrid generation is to produce hydrogen during off-peak periods or times when there is excess renewable electricity. An example is using wind power at night when wind speeds tend to be greater and electrical power requirements are often less. The hydrogen can be converted back to electricity to provide constant power when the renewable source is unavailable, helping with microgrid stabilization [24]. Hydrogen can be recombined with other chemical and compounds for other uses. For example, captured CO_2 can be combined to produce a synthetic natural gas that can be used in power plants or transportation applications [24].

THERMAL ENERGY STORAGE

Examples of thermal energy storage include molten salt energy storage, water, ice storage, or seasonal thermal storage. To increase functional capacities, concentrating solar power (CSP) technologies often use thermal energy storage. Common heat transfer fluids include synthetic oils, molten salt, and pressurized water. The fluids are used to transfer heat to water which changes state to steam to drive a steam turbine. These often use parabolic trough collector systems, one of the most tested and proven types of CSP technologies. Molten salt energy storage is most commonly used for solar thermal microgrid applications.

Built in three 50 MW phases, the Andasol Solar Power Station (150 MW) in Aldeire y La Calahorra, Spain, was the world's largest and first commercial-scale solar parabolic trough CSP power plant to combine parabolic trough solar CSP with molten salt energy storage (MSES) technology [25]. The plant is configured with solar arrays, thermal storage systems, heat exchangers, steam turbines, generators, and condensers plus dry cooling systems to reduce water consumption [25]. The net electrical capacity at the site is 150 GWh per year [25]. Able to store 7.5 hours of solar thermal energy, it achieves a plant capacity of 39% with a 95% system efficiency [25]. It was completed in 2011 at a cost of $1.24 billion.

Water also can be used directly as a heat storage mechanism without necessarily having to change state. The University of the Sunshine Coast (USC), Queensland, Australia, used a water battery system with over 6,000 solar panels and a thermal energy storage tank to reduce grid energy use at the Sunshine Coast campus by 40% [26]. The massive rooftop solar system constructed at USC is connected to a thermal energy storage battery with a thermal chiller and includes a real-time monitoring system [26]. When the sun shines, solar energy powers the thermal chiller to chill water which is stored in a 4.5-million-liter tank and used to air condition the entire campus [26]. The system delivers 2.1 MW of electricity and is expected to save more than $100 million in grid-supplied electricity costs over a 25-year period [27].

SUMMARY

Central problems with the use of electricity are that electricity must be used the moment it is produced and that it cannot be directly stored. When comparing generation system capacities, it is important to assess the ability of each type of storage system to store energy and to recapture stored energy [28]. This comparison should consider the instantaneous power, total energy production, and the degree to which energy can be stored and then dispatched from the storage system when needed [28]. As Whiting notes, stored energy "that can be generated when it is not needed, and then used when it is needed, makes it possible to convert energy generated by intermittent and variable sources to energy that can be dispatched as baseload" [28]. These are central issues that must be addressed when designing electrical energy storage systems for microgrids. Categories of storage technologies include electrical energy storage, mechanical energy storage, electrical/electrochemical energy storage, chemical and biological energy storage, and thermal energy storage. There are battery storage systems that use lead-acid, lithium-ion, sodium-sulfur, and vanadium redox technologies. Any of these technologies can be used to store power for microgrids.

Microgrids provide a ready market for energy storage. Unlike hydropower and geothermal generation resources, renewables such as solar and wind power are classic examples of resources that provide intermittent generation. To increase capacities, meet periods of high demand, and smooth the electric power provided to the system, energy storage is often a necessity. Reserve capacity is needed to maintain high-penetration microgrids [29]. There is a growing realization that electrical energy storage systems will be a key component of future electricity transmission networks, particularly those with heavy dependence on renewable resources [30]. Modern energy storage systems: 1) enable a match between supply and demand; 2) replace inefficient auxiliary power production; 3) ensure electric grid stability with a diversified energy supply and increased levels of renewable penetration; 4) ensure security of supply; and 5) facilitate distributed generation [30]. There are numerous types of electrical and thermal energy storage technologies, differentiated by power and energy density, physical size, cost, charge and discharge time periods, and market readiness (see Figures 7.3 and 7.4) [30].

Fundamentals of Microgrids

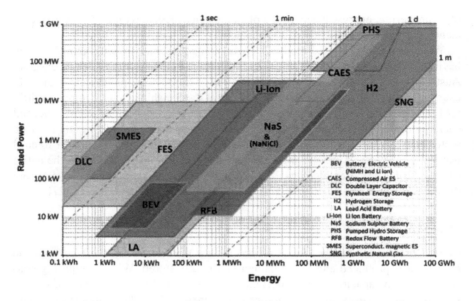

FIGURE 7.3 Comparative power, energy, and discharge durations for selected storage technologies. (Source: International Electrotechnical Commission, EES Technologies [22].)

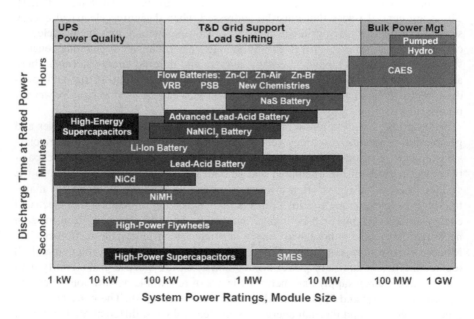

FIGURE 7.4 Positioning of energy storage technologies. (Source: EPRI, Sandia National Labs [23].)

Conversion losses are inevitable in any energy storage system charging and discharging cycle. Use of renewable energy in microgrids introduces the problem of intermittent generation. Fast-ramping generation and storage assets working in concert with the necessary control devices can rapidly compensate for intermittent generation [29]. Flywheels, batteries, or other storage technologies can serve as auxiliary generators and provide ride-through capability [29]. There are other types of energy storage systems that can provide longer duration storage for microgrids. These include mechanical energy storage systems that store energy such as compressed air or as higher elevation water as in pumped storage hydropower systems. In terms of total megawatts of installed storage capacity, pumped storage hydroelectricity is the primary technology used for microgrid applications.

REFERENCES

1. Microgrids at Berkeley Lab (2018). Microgrid definitions. https://building-microgrid.lbl.gov/microgrid-definitions, accessed 24 January 2020.
2. Carnegie, R., Gotham, D., Nderitu, D. and Prekel, P. (2013, June). Utility scale energy storage systems – benefits, applications and technologies. State Utility Forecasting Group. https://www.purdue.edu/discoverypark/energy/assets/pdfs/SUFG/publications/SUFG%20Energy%20Storage%20Report.pdf, accessed 19 May 2020.
3. International Electrotechnical Commission (2011). Electrical energy storage.
4. Sauty, F. and Tomlinson, C. (2016, February). Hybrid microgrids: the time is now. http://s7d2.scene7.com/is/content/Caterpillar/C10868274, accessed 24 January 2020.
5. Marine Microgrids. Aalborg University. https://www.et.aau.dk/research-programmes/microgrids/mission-and-focus-areas/maritime-microgrids, accessed 24 November 2018.
6. U.S. Energy Information Administration (2019, October 31). Most pumped storage electricity generators in the U.S. were built in the 1970s. https://www.eia.gov/todayinenergy/detail.php?id=41833, accessed 21 November 2019.
7. Andres, R. (2018, October 2). El Hierro third quarter 2018 performance update. http://euanmearns.com/el-hierro-third-quarter-2018-performance-update, accessed 1 December 2019.
8. Carpenter, K. (2018). Compressed air systems, in *Energy Management Handbook* (Roosa, S. ed.). Fairmont Press: Lilburn, Georgia, page 592.
9. Wikipedia. Compressed air energy storage. https://en.wikipedia.org/wiki/Compressed_air_energy_storage, accessed 7 December 2018.
10. Latha, L., Palanivel, S. and Kanakaraj, J. (2012). Frequency control of microgrid based on compressed air energy storage system. *Distributed Generation and Alternative Energy Journal*, 27(4), pages 8–9.
11. Energy Storage Association (2018). Compressed air energy storage (CAES). http://energystorage.org/compressed-air-energy-storage-caes, accessed 1 December 2018.
12. Hanson, G. (2019, December 20). What is flywheel energy storage? https://www.wisegeek.com/what-is-flywheel-energy-storage.htm, accessed 24 January 2020.
13. Breeze, P. (2018). *Power System Energy Storage Technologies. Academic Press*: New York, NY.
14. Research Interfaces (2018, April 14). Lithium-ion batteries for large-scale grid energy storage. https://researchinterfaces.com/lithium-ion-batteries-grid-energy-storage, accessed 6 December 2018.
15. Murdoch University. Stand-alone power supply systems. http://www.see.murdoch.edu.au/resources/info/Applic/SPS, accessed 20 December 2018.

16. International Renewable Energy Agency (2017). Electricity storage and renewables: costs and market to 2030. http://www.irena.org/publications/2017/Oct/Electricity-storage-and-renewables-costs-and-markets, accessed 6 December 2018.
17. Porter, G. and Sciver, J., editors (1999). *Power Quality Solutions: Case Studies for Troubleshooters*. Fairmont Press: Lilburn, Georgia, page 31.
18. Lithium ion rechargeable batteries. Technical handbook. http://www.sony.com.cn/products/ed/battery/download.pdf, accessed 6 December 2018.
19. Richmond, R. (2013, February). Lithium-ion batteries for off-grid systems – are they a good match? *Home Power*, page 153.
20. MaRS Cleantech. The future of microgrids. https://www.marsdd.com/wp-content/uploads/2016/10/Future-of-Microgrids-Residential.pdf, accessed 17 December 2018.
21. Davidson, P. (2007, July 5). New battery packs power punch. *USA Today*, page 3B.
22. International Electrotechnical Commission (2011). Electrical energy storage, white paper. Geneva, Switzerland. http://www.iec.ch/whitepaper/pdf/iecWP-energystorage-LRen.pdf, accessed 24 January 2020.
23. U.S. DOE/EPRI, Sandia National Labs (2013, July). Electricity storage handbook in collaboration with NRECA. https://www.sandia.gov/ess-ssl/lab_pubs/doeepri-electricity-storage-handbook, accessed 22 November 2019.
24. Thomas, J. (2013, April). Hydrogen: a promising fuel and energy storage solution. *Continuum Magazine*, 4, page 22.
25. Power Technology. The Andasol solar power station project. https://www.power-technology.com/projects/andasolsolarpower, accessed 22 November 2019.
26. May, J. (2019, November 27). Using renewables. Solar energy produced "water battery" wins global award and saves 40% of university's energy use. https://usingrenewables.blogspot.com/2019/11/solar-energy-produced-water-battery.html?spref=fb&fbclid=IwAR1ahyK21gTfWkSrF6mzlU0LPYqDciAHvSF2J0ET2rZPuegRscyboKVFnQk, accessed 27 November 2019.
27. University of the Sunshine Coast (2019, October). Sunny approach to cooling campus wins global award. https://www.usc.edu.au/explore/usc-news-exchange/news-archive/2019/october/sunny-approach-to-cooling-campus-wins-global-award, accessed 27 November 2019.
28. Whiting, G. (2019). Electric generation capacity and 10% renewable energy goals. *International Journal of Energy Management*, 1(3), pages 8–25.
29. GTM Research (2016, February). Integrating high levels of renewables into microgrids: opportunities, challenges and strategies. www.sustainablepowersystems.com/wp-content/uploads/2016/03/GTM-Whitepaper-Integrating-High-Levels-of-Renewables-into-Microgrids.pdf, accessed 16 January 2019.

8 Hybrid Generation Systems for Microgrids

ENERGY RESOURCES FOR MICROGRIDS

Most energy resources available can be used to generate electricity for microgrids. Despite the empowering benefits of local control over generation, there are downsides associated with the use of standalone generation that are specific to the technologies deployed. This is true for both fossil fuel-fired and renewable generation. Diesel fuel has pollution and supply chain issues. Solar power cannot be directly produced at night. Wind power has intermittency issues associated with the wind's variability at any given location. Such issues must be addressed and overcome when microgrids are used for power generation. How can these problems be resolved? Combing different types of generation is common approach.

Utility companies and independent power producers are seeking standardized microgrid models that offer a wider range of benefits for consumers [1]. There is a need for power reliability and scalable plug-and-play electrical generation systems. In a push for more scalability and improved sustainability, more electric utilities are considering hybrid generation microgrid models [1]. An important feature of microgrids is the mix of energy resources often used for electrical generation. Many microgrids are being configured as hybrid generation systems, those that use multiple technologies to achieve their resiliency goals. Having multiple forms of generation offers microgrid developers flexibility in microgrid design and configuration. This morphological convergence is a reason why hybrid electricity generation technologies are often considered in the design of modern microgrids.

Distributed energy generation technologies are often interpreted as sustainable system innovations that support the emergence of a new technical regime by promoting creativity and ultimately providing sustainable power generation [2]. Past research has proven that energy use, local actions, and environmental policies are closely linked to the concept of sustainable development [3]. Sustainability rankings of key electricity generation technologies indicate that distributed energy generation technologies, renewable-based electricity generation technologies, and combined heat and power (CHP) plants are more congruent with the principles of sustainable development than their conventional, large-scale counterparts [2].

From the perspective of sustainability advocates, a clean energy economy powered by both renewables and energy efficiency would be considered the most sustainable energy planning scenario available [4]. Traditionally, microgrids have served a single entity, such as a university or hospital, and have been powered by one source of fuel. They were often built using existing infrastructure and equipment [5]. Microgrids also allow for the use of clean and efficient distributed energy resources such as storage, wind, solar, and combined heat and power, often in combination [5].

These networks that provide multiple forms of generation and have internal distribution are hybrid microgrids. They are characterized by having more than one form of renewable generation, various forms of energy storage, and a central center for monitoring, communications and control. Locations with electrical loads often have generation or energy storage systems on-site (see Figure 8.1).

HYBRID GENERATION SYSTEMS

Like larger utilities that provide electrical power, microgrids must provide for the needs of their customers by providing electricity that meets fluctuating baseloads and variable electrical demand. These loads can fluctuate hourly or daily and are subject to seasonal variability. This challenge can be met by using a strategic mix of generation resources. Having multiple generation technologies available allows microgrids to function under various conditions of electrical loads and provide a more resilient operation.

What Are Hybrid Power Systems?

Hybrid electrical systems attempt to achieve greater independence from the central grid by using multiple generation systems. *Hybrid energy systems* (HES) combine two or more energy conversion mechanisms, or two or more fuels for the same mechanism to generate electricity or power [6]. When integrated, they have the potential to overcome limitations inherent in either, such as balancing power or increasing system efficiency [6]. A hybrid renewable energy system (HRES) uses two or more types of renewable energy to generate electricity or power. For hybrid systems, energy storage systems are both an electricity power source and a load.

Hybrid renewable energy systems are usually (but not always) labeled by naming the dominate generation system first. For example, a *solar thermal hybrid generation system* would refer to a HES that predominately generated electricity using solar thermal energy coupled with a less dominate form of generation, such as diesel gensets. In this type of configuration, there is no question that this represents a HES as there is clearly more than a single type of generation technology being deployed.

A question that seems to be unresolved is whether or not a system that uses one type of generation (e.g., solar) combined with one form of battery storage (e.g., sodium-sulfur or lithium-ion) can be defined as a hybrid system. The argument in favor of categorizing such a system as a hybrid alludes to that fact that when batteries are used, they can be considered to be a form of generation when discharging energy. The argument against calling such configurations hybrid systems is that there are not two technologies used for direct electrical power generation and that batteries are used as a storage medium only. As the use of battery storage systems expands, the first argument is becoming more widely adopted.

Typically, a hybrid system might contain alternating current (AC) diesel generators, direct current (DC) diesel generators, a compatible distribution system, loads, renewable power sources (wind turbines, biomass and hydroelectric generators, geothermal generation, or photovoltaic power sources), energy storage, power converters, rotary converters, coupled diesel systems, dump loads, load management

Hybrid Generation Systems for Microgrids

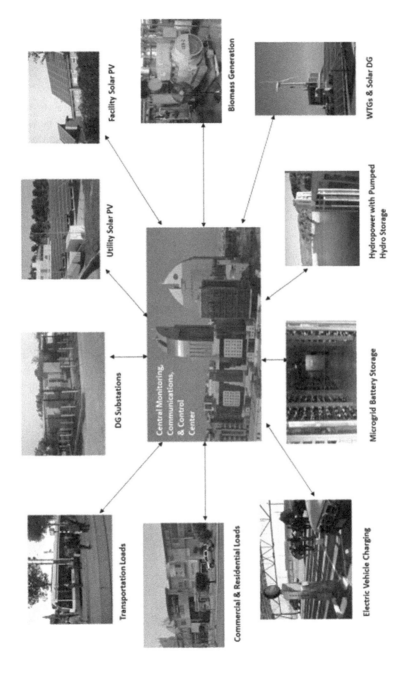

FIGURE 8.1 Microgrid configuration using hybrid distributed generation systems and energy storage.

options, or a supervisory control system [7]. Larger systems (>100 kW), typically consist of AC-connected diesel generators, renewable generation sources, and loads, and occasionally include energy storage subsystems [7]. Below 100 kW, combinations of both AC- and DC-connected components with energy storage are common [7].

Advantages of Hybrid Generation

Hybrid generation systems are seen as a solution by utility companies to achieve scalability and standardization of microgrid models that maximize the benefits for both companies and end users [8]. Power reliability is an important goal in the design and development of hybrid generation systems. These systems must efficiently and economically satisfy the loads for the microgrid to be workable.

Hybrid power generation systems can be used for remote area power system (RAPS) applications. According to Electropedia, certain hybrid renewable power generation systems (e.g., solar and wind power combinations) have the following advantages [9]:

Complementary generation: the renewable electrical generation systems can be configured to be complementary. For example, during the summer months when there is a lack of wind there is often ample sunlight. During winter months there is less sunlight, but stronger winds are more prevalent.
Supply diversity: two or more different energy sources provide a diversity of supply, reducing the risk that power outages might occur.
Reduced equipment costs: high cost ancillary equipment (e.g., batteries and inverter) that are required for a single system can be specified to carry the full system load. A second system can be added without increasing its capacity or adding cost for more of these components.
Lower energy storage requirements: because of the supply diversity, the required capacity of an energy storage system can likely be reduced to lower the cost of the total system.
Loads sharing: the required generating capacity of solar and wind energy conversion units can be reduced since the total load is shared [9].

To assess the feasibility of hybrid systems the comparative advantages and disadvantages of the various generation systems considered must be weighed against their costs for installation, maintenance, and operation. This can be a complex undertaking as hybrid systems can be configured in innumerable ways. The various combinations might include any of the following:

Fossil fuel generation: internal combustion engines, Rankin cycle engines, gas turbines, Stirling engines, microturbines, and others.
Renewable energy generation: solar photovoltaics, solar thermal, concentrating solar power, wind turbine generators, hydropower, biomass combustion systems, geothermal flash and binary technologies, and others.

Fuel cells: solid oxide, proton exchange membrane, phosphoric acid, molten carbonate, etc.

Storage: high-power supercapacitors, flywheels, flow batteries, lead-acid batteries, lithium-ion batteries, sodium-sulfur batteries, reversable fuel batteries, thermal storage, compressed air, pumped hydropower, biofuels, hydrogen, etc.

Control systems: the system hardware and software components, sensing systems, functions, and algorithmic capabilities that orchestrate operations and enhance efficiencies.

EXAMPLES OF HYBRID GENERATION

Hybrid microgrids are becoming a preferred approach to delivering cost-effective, reliable power in locations beyond the reach of larger electric utility infrastructure [10]. There are numerous templates for hybrid electrical generation systems that can be used for microgrid applications. Typical models combine fossil fuel generation, one or more forms of renewable energy, and energy storage in a configuration with a microgrid control system. Examples of the types of combinations are described in this section.

DIESEL AND RENEWABLE HYBRID SYSTEMS

Remote locations such as mines, drilling sites, or harvesting centers are beyond connections to national electric grids and often employ diesel generation for electric power. These are typically small-scale microgrids. Fuel must be delivered to the generator which can be costly due to the remote location of such sites. When the diesel gensets are combined with renewable energy sources such as wind turbines or solar panels we refer to them as hybrid systems. With hybrid systems two goals are important—reducing fuel use and maintaining system reliability [11].

Hydro, wind, or solar power can reduce the reliance on power produced from generator sets, reducing fuel consumption and maintenance costs while offsetting weaknesses [10]. The diesel generators firm the renewable sources and follow the loads. Sophisticated direct digital controls (DDC) link the generation systems. Energy storage enhances systems helping the generator sets respond smoothly to significant fluctuations in output from the renewable resources, while maintaining consistent voltage and frequency [10].

NATURAL GAS AND RENEWABLE ENERGY SYSTEMS

Advances in technologies supporting natural gas-fired steam turbines have led to their increased use in hybrid electrical generation systems, linking them with renewable energy generation (see Figure 8.2). These advances, including digital trip systems, electronic speed governors, and software capabilities, have provided improvements in monitoring, safety, reliability, efficiency, and performance [12]. Today's advanced steam turbine systems are used for direct power generation, district heating, and

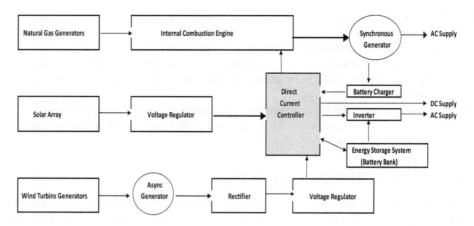

FIGURE 8.2 Microgrid configuration using both natural gas, solar PV, and WTGs. (Adapted from [9].)

seawater desalinization, and can be combined with waste-to-energy, biomass, and solar thermal generation [12].

Decentralized applications that combine heat and power systems with renewables and storage create more resilient generation systems capable of providing continuous electrical generation for remote locations and businesses seeking to lower costs. Oil fields in remote areas of Peru exhaust natural gas which is relieved into the atmosphere or flared. An interesting solution has been proposed to optimize the oil-field production facilities' electrical system and lower costs which involves combining wind power with natural gas to provide electricity [13]. The project proved that the use of a hybrid wind–thermal natural gas system to generate electricity would enable a reduction in the use of fuel gas that is required to supply electricity for field operations [13].

SOLAR AND WIND POWER

A common configuration for microgrids is to combine solar PV with wind turbine generators (WTGs). The logic is that solar energy is abundant during the daylight hours and that winds tend to be stronger at night, especially over land areas of the northern hemisphere. Often this combination uses some kind of battery storage. A micro-controller can be used to ensure optimization of linked solar and wind generation systems and improve their efficiency.

A waste water treatment facility serving Atlantic City, New Jersey has a hybrid solar-wind power plant that generates over 40.8 million kWh of electricity annually (see Figure 8.3). The plant has an all-electric fleet of support vehicles that are recharged on-site using electricity generated by the hybrid system. Excess electricity is sold back to the electric grid powering the nearby casinos.

There are numerous microgrids that use these generation sources that are planned or under construction. EnSync Energy Systems announced a partnership with WindStax Energy to develop a microgrid for a new trucking terminal in Parma, Ohio

Hybrid Generation Systems for Microgrids

FIGURE 8.3 Hybrid wind and solar PV generation for the Atlantic City wastewater plant.

in 2019 [14]. The microgrid will use a 495 kW solar PV system, have eight WTGs that add 48 kW, and also feature 760 kW of energy storage [14].

Prior to ramping up to full operation, microgrids are tested. Ameren Corporation and S&C Electric Co. recently proved that their automated microgrid can successfully island and provide utility-scale power without help from the larger electric grid [15]. In Champaign, Illinois, the companies used a 50 kW installation to test the microgrid's capabilities by supplying power to an Ameren research facility for a 24-hour period [15]. The entire microgrid includes 225 kW of solar and wind generation plus 500 kW of battery energy storage [15]. By powering the facility for the duration of the testing, the microgrid proved that it is capable of operating at utility-scale voltages and can seamlessly transition from grid-connected to islanded modes [15].

SOLAR AND GEOTHERMAL ENERGY

A project in Nevada's Great Basin is unique in that it offers an example of combining a medium enthalpy, binary cycle geothermal plant with solar PV and solar thermal to generate electricity—a triple generation plant (see Figure 8.4). It was the world's first plant to couple these three forms of generation technologies. The first phase of the Stillwater Project began with a 33.1 MW geothermal binary plant which was commissioned in 2009 [16]. In 2013, a 26.4 MW solar project using concentrated solar power (CSP) thermal technology was initiated on the site. The augmentation project relies on linear parabolic trough systems to add energy to the incoming geothermal fluid, which allows the binary plant to increase output [16]. The project includes collectors, a heat exchanger, circulating pumps, and a control system integrated with the geothermal plant [16]. The solar arrays add about 17 MW of thermal energy, plus an

FIGURE 8.4 ENEL Green Power North America's Stillwater plant in Fallon, NV. (Sources: Idaho National Laboratory and ENEL Green Power of NA [17].)

equivalent of 2 MW of boost in power generation to the geothermal power plant [16]. The plant generates enough electricity to power 15,000 residences [17].

Using a combination of technologies with different generation profiles increases energy availability and reduces intermittency [17]. Geothermal (baseload capable) and solar (an intermittent resource) power are complementary technologies when used for electricity generation. The geothermal plant pumps heated geothermal brine and transfers the heat to a secondary fluid, isobutane, which flashes into a steam-like substance spinning the turbines as it expands [17]. Having yielded its heat, the brine is circulated back into the Earth while the isobutane cools, condenses into a liquid state, and begins another boil-and-expand cycle [17]. For the solar thermal process, the sun's rays focus onto tubes of demineralized water [17]. The heat from this water is transferred to the incoming geothermal brine, increasing the amount of energy available to boil the isobutane and creating more power to spin the turbines that produce electricity [17]. The plant's use of air as a cooling medium is important as water shortages have intensified throughout parts of the western U.S. [17]. In addition, there are 240 acres (97 hectares) of adjacent solar PV panels that work in tandem with the geothermal system [17].

The complementary nature of the hybrid generation technologies used at the Stillwater plant support the project's cash flow by improving efficiency and reducing production costs. The project offers an example of how renewable technologies can work in tandem to optimize generation. Production from solar is higher during the sunniest and hottest days of the year, when the geothermal plant's thermal efficiency is lower [18]. During the sunniest part of the day, when the geothermal system's output declines, the solar PV output is at its maximum [17]. Likewise, when the geothermal energy output strengthens during the cool desert nights, the solar PV system is unable to generate electricity due to lack of sunshine [17]. The increased electricity production during peak periods enables a load profile that better matches production to loads.

The strategy to combine renewable sources at a single site utilizes previously installed assets which maximize the project's return on investment. Combining several renewable power generation technologies increases output without increasing emission or environmental impacts by sharing the existing electrical interconnection substation, transmission lines, access roads, control building, and other common facilities [16]. Post-completion research has determined that by combing the hybrid generation technologies the overall output at the Stillwater plant was increased by 3.6% [18].

Nuclear and Renewable Energy

The main purpose of a nuclear-renewable hybrid energy system (N-R HES) is to use nuclear energy in combination with variable renewable energy sources such as wind and solar, biomass energy, or others to support the electrical and thermal requirements of electricity generation, fuels production, and chemical synthesis [19]. Synergistic benefits of these systems might include reducing costs, limiting water use, maximizing dispatchable power, improved grid support, or more efficient utilization of capital [20]. The design of the nuclear reactor could be sized and optimized for the particular industrial microgrid that it serves. It is conceivable that combining these dissimilar systems to support process applications can reduce costs and decrease greenhouse gas emissions (GHG) in the manufacturing, industrial, and transportation sectors [19].

The major components of an N-R HES system would include nuclear reactors with steam generation, renewable power generation, industrial process loads, and storage capabilities. Flexible generation often has greater value in electricity markets than baseload or variable supplies [19]. Flexible N-R HES architectures include the following [20]:

Tightly coupled N-R HES: in this architecture, nuclear and renewable generation sources and industrial processes would be linked and co-controlled behind the electricity bus, such that there would only be a single grid connection. The closely coupled system would be managed by a single entity to optimize profitability for the integrated system [20].

Thermally coupled N-R HES: this architecture would thermally integrate subsystems and tightly couple them to the industrial processes, but the nuclear and renewable electrical subsystems could have multiple connections to the same grid balancing area and would not need to be co-located. They would be centrally controlled to provide energy and ancillary services to the grid. The thermally integrated subsystems would need to meet industrial process requirements considering the required heat quality, the heat losses to the environment, and the required exclusion zone around the nuclear plant. These systems would likely be managed by a single financial entity [20].

Hybrid N-R HES: this configuration would be electrically coupled to industrial energy users but lack direct thermal coupling of subsystems. This design would allow management of the electricity produced within the system (e.g., from the nuclear plant or from renewable electricity generation) prior to the grid connection. Although there would be no direct coupling of thermal energy to the industrial processes, the system could include electrical to thermal energy conversion equipment to provide thermal energy input to the industrial processes. These systems might have multiple connection points to the grid. Molten salt reservoirs could provide thermal energy storage to generate electricity. Electrical-to-thermal energy conversion becomes economical when the cost of producing heat drops below the cost of producing heat from combustion-fired process heaters [20].

The design and configuration of the hybrid options that combine nuclear power with renewable energy resources would vary based on loads, efficiencies, availability of water, and types of renewable generation and storage systems. For example, it is possible to design the plant to generate clean hydrogen with any excess electricity using a steam electrolysis plant.

Fuel Cells and Renewable Hybrid Generation

Using fuel cell technologies within microgrids is a complementary use of the technologies. Fuel cell use in microgrid configurations is growing as applications are found that are economically feasible particularly for stationary applications. About 80% of Bloom Energy's deployments of fuel cells are for microgrid applications [21]. Microgrids using fuel cells can readily separate from the grid without impacting reliability and the quality of electricity [21]. This capability plus their ability to quickly generate electricity makes them compatible with hybrid generation systems that use renewable energy.

Fuel cells are a reliable clean energy source that is modular, scalable, and emissions-free. Fuel cells are not as cost-effective as most forms of conventional generation but they are suitable for making use of renewable energy and can be incorporated with community-scale energy systems [22]. Other benefits of fuel cells include low noise levels during operation and a relatively small installation footprint [23]. Despite these merits, there were only 116 MW of fuel cells deployed to support microgrids by 2016 [23].

Fuel cells are similar to batteries. They generate electricity from an electrochemical reaction using hydrogen and require a constant fuel supply [23]. Fuel cells can operate and generate electricity so long as the fuel, normally hydrogen, is available. While hydrogen is abundant, it is usually produced by reforming fossil fuels such as natural gas, methane, or butane. Renewable hydrogen can be produced using a renewable energy source and electrolysis to split water into its elemental components. Though the efficiency of fuel cells is much better that of gas-fired turbines and they produce only half the emissions, they are much more expensive than gas turbines on a per watt basis [23].

Examples of fuel cell-only microgrids include the University of Bridgeport, in Connecticut, which is supplied electricity by an investor-owned fuel cell power plant. It is configured for baseload power, net metering, and CHP. A sample of the many examples of fuel cells being linked with hybrid renewable energy systems are noted below.

The *University of California San Diego* microgrid produces 92% of its electricity and 95% of its heating and cooling loads using a fuel cell powered by directed biogas. The multi-generator system has a fuel cell (2.8 MW), a roof-mounted solar PV (3 MW) system, a steam turbine (3 MW), plus two gas turbines (13.5 MW each) [24]. Electricity is provided to the campus 12 kV distribution system [24].

The *Gordon Bubolz Nature Preserve* in Appleton, Wisconsin is powered by a microgrid that includes solar PV (200 kW), a hydrogen fuel cell (30 kW), a micro-turbine (65 kW), a natural gas generator (60 kW), and a lithium-ion battery storage system (100 kW) [25]. The microgrid's integrated distributed energy resources (DERs) are managed by an energy control center using a software platform that autonomously configures the DERs to optimize the efficiency and cost-effectiveness of the combination of energy resources being used at any given time [25].

The *Santa Rita Jail*, a microgrid demonstration project in Dublin, California, is configured with five wind turbine generators (2.3 kW), a solar PV array (1.2 MW), two diesel generators (1.2 MW), a fuel cell (1 MW), and advanced energy storage (2 MW) [26]. It has a DER management system which reduces peak demand during normal grid-connected operation or during demand response events [26].

Using the methane emitted from landfills reduces GHG emissions. In Riverside, California methane gas from a local landfill is used to power fuel cells [21]. Wastewater treatment plants, dairies, and agricultural processing plants offer additional opportunities to use methane for biogas-fueled fuel cells [21].

Renewable Co-Generation

Co-generation normally refers to the process by which electricity is generated on-site by a user to displace electrical energy purchased from a utility [27]. Historically, the requisites for co-generation are: 1) high electric demand with heavy electrical consumption; 2) a suitably large thermal load; and 3) high demand and/or unit energy costs [27]. Benefits of these systems include reduced heat losses and higher efficiency.

Though primarily considered when using fossil fuel generation, co-generation is also feasible in renewable energy applications. Examples of applications include CHP systems that are based on biofuels, solar, or geothermal energy. Trigeneration refers to the simultaneous generation of electricity plus useful heating and cooling from a combustion process or a solar heat collector [28]. Hybrid renewable energy systems can be used on a dedicated site to provide renewable co-generation or trigeneration. In a microgrid configuration, two solar plants, one thermal and a separate solar PV array, can be used to provide combined heat and power. Solar PV combined with a fuel cell can be configured to provide micro-combined heat and power systems.

Renewable co-generation can be achieved with the use of a biogas created from organic wastes. Biologically derived biogases (also called biomethanation) are produced as metabolic products of microorganisms called bacteria and archaea [29]. Biogases are waste products from agricultural processes and are composed of methane gas and carbon dioxide with trace quantities of hydrogen or nitrogen. The waste gases can be produced by using anaerobic fermentation processes. These gases can be used as a renewable power source in biogas co-generation systems [29]. Anaerobic digestion can occur at mesophilic (35–45°C) or thermophilic temperatures (50–60°C) but supplementary sources of heat are required to reach their optimal temperatures [29]. This heat can be provided by a biogas CHP unit as a standalone system or supplemented with solar thermal heat.

Waste streams might include cattle dung which produces high concentrations of methane gas or waste from municipal sewage. These biogas engines are linked to an alternator to produce high efficiency electricity which enables end-users to maximize the electrical output of the system and optimize the plant's economic performance [29]. The benefits of using biogas produced by this process include: 1) production of renewable power using CHP; 2) disposal of problematic waste products; 3) diversion of waste from landfills; 4) production of a low-cost, low-carbon fertilizer; 5) avoidance of landfill and feedlot gas escaping to the atmosphere; and 6) reduction of carbon emissions [29].

SUMMARY

Hybrid energy generation systems are commonly used in microgrids. They generate electrical power by relying on the use of two or more types of generation technologies. To be effective, these systems must overcome the inherent limitations of providing electricity with a single energy source. While there are many potential combinations of hybrid generation systems that have been used in microgrid deployment, this chapter detailed the examples that included diesel and renewable hybrid systems, natural gas and renewable energy systems, solar and wind power, solar and geothermal energy, nuclear and renewable energy, fuel cells, and renewable hybrid generation.

Hybrid systems can be used to generate AC or DC power or both. Many renewable energy systems natively generate DC power. When coupled with DC electrical loads, equipment costs can be reduced. Figure 8.5 graphically depicts the typical configuration of hybrid generation systems using fossil fuels and renewables. Renewable co-generation and trigeneration systems were also considered. Trigeneration (providing cooling, heat, and power) is a form of hybrid generation and refers to the simultaneous generation of electricity and useful heating and cooling from a combustion process or a renewable energy source.

Energy storage is often coupled with hybrid generation systems. Since electricity cannot be directly stored, energy storage systems suffer from energy losses due to conversion losses. There is debate in the industry as to whether or not an electrical generation system that uses battery storage in conjunction with a single generation technology constitutes a hybrid system.

To assess the feasibility of hybrid systems the comparative advantages and disadvantages of the various alternative combinations of generation systems must be considered. Using two or more different energy sources provides a diversity of supply, reduces the risk that power outages might occur, and improves overall system reliability and resiliency. The benefits of co-locating multiple renewable generation types at or near a single facility include having access to more accurate resource and

Hybrid Generation Systems for Microgrids

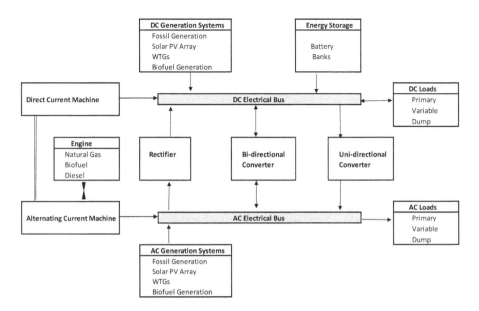

FIGURE 8.5 Hybrid configuration using fossil fuels and renewables. (Adapted from [7].)

environmental data, reductions in shared equipment, lower operations and maintenance costs, and the efficient harnessing of resources such as institutional capital or a knowledgeable team of available professionals [16].

Providing microgrid power with multiple generation technologies increases cost. These costs must be outweighed by the benefits of the services provided. There are three value propositions that overlap, increasing the value of hybrid systems for microgrids but also making the control systems more complex [30]. The market is being driven by a combination of: 1) resilience and reliability concerns; 2) sustainability and climate concerns increasing the rate of solar PV adoption; and 3) the decline in costs for batteries and other forms of energy storage that can be used for demand charge reduction [30].

The clean energy economy is powered by both renewables and energy efficiency. Microgrids use hybrid generation to increase the use of renewables, improve generation efficiencies, and reduce GHG emissions. In an effort to deploy electrical generation models that are more scalable and sustainable, today's electric utilities are exploring the use of hybrid microgrids. Hybrid generation systems have been successfully deployed in military, community, and industrial microgrids.

REFERENCES

1. Trabish, H. (2017, February 23). Utilities' microgrid strategy. *New Energy News.* http://newenergynews.blogspot.com/2017/08/original-reporting-utilities-microgrid.html, accessed 2 December 2018.
2. Deutsch, N. (2018). Electricity generation technologies and sustainability: the case of decentralized generation. *WSEAS Transactions on Business and Economics*, 15, pages 157–158.

3. Roosa, S., editor (2010). *The Sustainable Development Handbook*, 2nd edition. The Fairmont Press: Lilburn, Georgia.
4. North Carolina Sustainable Energy Association (2018). What is clean energy? https://energync.org/what-is-clean-energy, accessed 25 November 2018.
5. New York State (2018). Microgrids 101. https://www.nyserda.ny.gov/All-Programs/Programs/NY-Prize/Resources-for-applicants/Microgrids-101, accessed 19 December 2019.
6. Singh, D. A seminar on hybrid power system. https://www.scribd.com/doc/126331906/Hybrid-Power-System-Ppt, accessed 3 December 2018.
7. CRES. Definition of hybrid power systems. http://www.cres.gr/hypos/files/fs_inferior01_h_files/definition_of_hps.htm, accessed 20 December 2019.
8. Trabish, H. (2017, February 23). Utilities' microgrid strategy. *New Energy News*. http://newenergynews.blogspot.com/2017/08/original-reporting-utilities-microgrid.html, accessed 26 November 2018.
9. Electropedia, Woodbank Communications (2005). Battery and energy technologies, hybrid power generation systems. https://www.mpoweruk.com/hybrid_power.htm, accessed 27 November 2018.
10. Saury, F. and Tomlinson, C. (2016, February). Hybrid microgrids: the time is now. http://s7d2.scene7.com/is/content/Caterpillar/C10868274, accessed 24 January 2020.
11. ComAp. The heart of smart control for hybrid power generation, hybrid – the future of microgrids. https://www.comap-control.com/solutions/technology/hybrid, accessed 22 December 2019.
12. Brzozowski, C. (2018, November). Achieving unprecedented efficiency. *Distributed Energy*, pages 32–37.
13. Porles, F. (2018). Hybrid electrical generation for grid independent oil and gas well fields, in *International Solution to Sustainable Energy, Policies and Applications* (Roosa, S., ed.), The Fairmont Press: Lilburn, Georgia.
14. Pickerel, K. (2018, December 4). Microgrid for Ohio trucking terminal. *Solar Power World*. https://www.solarpowerworldonline.com/2018/12/ensync-is-building-a-solarwindstorage-microgrid-for-ohio-trucking-terminal, accessed 20 December 2019.
15. Meehan, C. (2017, August 18). Solar, wind-powered microgrid proves ready to power local grid in Illinois. https://www.solarreviews.com/news/solar-wind-microgrid-powers-grid-illinois-081817, accessed 20 December 2019.
16. DiMarzio, G., Angelini, L., Price, W., Chin, C., and Harris, S. (2015, April 19-25). The Stillwater triple hybrid power plant: integrating geothermal, solar photovoltaic and solar thermal power generation. *Proceedings World Geothermal Congress* 2015. Melbourne, Australia. https://pangea.stanford.edu/ERE/db/WGC/papers/WGC/2015/38001.pdf, accessed 1 December 2019.
17. Idaho National Laboratory. Hybrid power plant boosts by combining three energy sources. https://inl.gov/article/integrating-renewable-energy, accessed 3 December 2019.
18. ENEL Green Power (2016, March 29). ENEL Green Power inaugurates triple renewable hybrid plant in U.S. https://www.enelgreenpower.com/media/press/d/2016/03/enel-green-power-inaugurates-triple-renewable-hybrid-plant-in-the-us, accessed 1 December 2019.
19. U.S. Department of Energy. Quadrennial Technology Review 2015, chapter 4, Advancing clean electric power technologies. https://www.energy.gov/sites/prod/files/2016/06/f32/QTR2015-4K-Hybrid-Nuclear-Renewable-Energy-Systems.pdf, accessed 13 December 2019.
20. Bragg-Sitton, S., Boardman, R., Rabiti, C., Kim, J., McKellar, M., Sabharwall, P., Chen, J., Cetiner, M., Harrison, T., and Qualls, A. (2016, March). Nuclear-renewable hybrid energy systems: 2016 technology development program plan. INL/EXT-16-38165, Idaho National Laboratory, Idaho Falls.

21. Egan, J. (2019, July). Fuel cells, no longer a fuel of tomorrow. *Distributed Energy*, pages 16–19.
22. Zsebik, A. and Noval, D. (2019). Integration of fuel cells with energy systems. *International Journal of Energy Management*, 1(1), page 56.
23. Maloney, P. (2016, May 10). Fuel cells are a good partner for microgrids, but costs limit deployment. https://www.utilitydive.com/news/fuel-cells-are-a-good-partner-for-microgrids-but-costs-limit-deployment/418891, accessed 20 December 2019.
24. Milliken, J. (2017, August 29). Fuel cells for microgrids, presentation. Northeast Electrochemical Energy Storage Cluster. http://neesc.org/wp-content/uploads/2017/08/Fc-for-microgrids-_binder_9-29-17v2.pdf, accessed 21 December 2019.
25. Power Engineering (2018, June 4). Wisconsin microgrid combines solar, wind, gas, battery storage and hydrogen fuel cell. https://www.power-eng.com/articles/2018/06/wisconsin-microgrid-combines-solar-wind-gas-battery-storage-and-hydrogen-fuel-cell.html, accessed 17 December 2019.
26. Ton, D. and Smith, M. (2012). The U.S. Department of Energy's microgrid initiative. https://www.energy.gov/sites/prod/files/2016/06/f32/The%20US%20Department%20of%20Energy's%20Microgrid%20Initiative.pdf, accessed 1 December 2019.
27. Payne, W., editor (1997). *Cogeneration Management Reference Guide*. The Fairmont Press: Lilburn, Georgia.
28. Wikipedia. Cogeneration. https://en.wikipedia.org/wiki/Cogeneration, accessed 24 November 2019.
29. Clarke Energy. Biogas. https://www.clarke-energy.com/biogas, accessed 20 December 2019.
30. Sanchez, L. (2019, June 24). A new era for microgrids (quoting Lilienthal, P.). *Distributed Energy*. https://www.distributedenergy.com/microgrids/article/21086075/a-new-era-for-microgrids, accessed 11 November 2019.

9 Community and Local Microgrids

Today, the majority of active microgrids provide electricity and sometimes heat to a single customer or entity. Many of these microgrids operate using fossil fuels, renewables, or a combination of both. Military bases and hospitals that require reliable sources of emergency power in the event of grid failures are among the many examples of contemporary microgrid-capable facilities. On a larger scale, microgrids allow communities to be more energy independent and in some cases provide lower cost electricity that is more environmentally friendly [1]. Community microgrids serve a targeted group of customers, such as municipal or public facilities, as hosts or tenants [2].

COMMUNITY MICROGRIDS

Community microgrids typically cross public rights of way and serve to modernize local utility infrastructure, providing resilient power for vital community assets [2]. Community microgrids come in a variety of designs and sizes but are generally integrated into utility networks. A community microgrid can power a single facility like the Santa Rita Jail microgrid in Dublin, California or power larger groups of connected buildings. For example, in Fort Collins, Colorado, a microgrid is part of a larger goal to create an entire district that produces all the energy that it consumes [1].

A *community microgrid* is a coordinated local grid area served by one or more distribution substations and supported by high penetrations of local renewables and distributed energy resource (DERs) such as energy storage [3]. Community microgrids are a new approach to designing and operating electric grids. They rely heavily on DERs to achieve a more sustainable, secure, and cost-effective energy delivery system while using power generated by renewables for prioritized loads over indefinite durations [3]. Key features of community-scale microgrids include: 1) high penetrations of renewable generation and DERs; 2) efficient load design which includes load balancing and flattening capabilities to reduce peaks and transmission costs; 3) ability to island critical loads; and 4) a scalable solution which spans one or electrical more substations [3].

DRIVERS FOR COMMUNITY-SCALE MICROGRIDS

Community microgrids are developed to support specific local goals and meet local requirements and mandates. They may also be used to reduce costs, or connect to a local resource that is too small or unreliable for traditional grid use. They are used for off-grid and grid-connected applications.

There is a need for microgrid deployment in the world's developing and developed countries especially for off-grid situations and village-scale community electrification programs. For developing countries, providing and maintaining energy access is an important driver for off-grid renewable energy systems that are often the most economical solution for providing electricity [4]. For developed countries, off-grid systems consist of two types: 1) mini-grids for rural communities, institutional buildings, and commercial/industrial plants and buildings; and 2) self-consumption of solar power generation for residences [4].

Mini-grids are particularly relevant for island states. There are more than 10,000 inhabited islands around the world and an estimated 750 million islanders [4]. Many of these islands, especially those with fewer than 100,000 inhabitants, rely on diesel generators for their electricity production and expend considerable funds on imported fuels [4]. Those that are tourist destinations are particularly sensitive about the pollution caused by fossil fuel generation. It is easier to locate clean energy generation near population centers from an environmental perspective. Due to delivery costs for diesel to remote locations, renewable generation is often a cost-effective replacement for diesel generation which creates market opportunities for off-grid renewable energy systems [4].

There are a number of drivers for the development of community microgrids. As they are already a conventional form of emergency or backup power for individual buildings and campuses, expanding existing microgrids can provide emergency backup generation for entire communities. Fossil fuel-fired central power stations that are located in or near cities are notorious point sources of pollution especially in developing countries. Restricting the use or eliminating these central stations and replacing them with renewable energy microgrids is a cause for the development of community microgrids. The present vulnerabilities of the utility grid create yet another driver to consider independent local power systems as an alternative. Reducing the impacts from power outages are a reason that municipal leaders consider enabling energy production closer to cities by using microgrids.

Cities large and small are taking initiatives to tap local energy resources and build energy infrastructure that will expand their tax base, diversify energy supplies, and address the need for expanded services for denser urban populations. Cities are seeing dramatic population growth along with a need for energy master planning and integrated energy systems that optimize efficiency and reduce waste [5].

U.S. Municipal Renewable Energy Goals

Over 90 U.S. cities, ten counties, and two states have made commitments to use 100% clean or renewable energy by 2050 [6]. According to the Sierra Club, "a community is powered with 100% renewable energy when the amount of energy generated from renewable energy sources in the community (or brought into it) equals or exceeds 100% of the annual energy consumed within the community" [6]. This definition supports the net zero community power concept. To achieve this goal, nuclear, natural gas, coal, oil, or other forms of carbon-based energy are not included as renewable energy sources.

There are a few cities that have already met their renewable energy goals. Examples include Aspen, Colorado; Rock Port, Missouri; Georgetown, Texas; Kodiak Island; and Burlington, Vermont [6]. Other larger cities (e.g., Orlando, Los Angeles, San Francisco, Saint Louis) have also set such goals within time frames of their choosing. Goals are being established that identify the total (often a percentage) of the community's energy supplies that will be met locally using distributed generation.

However, due to the costs of replacing the existing fossil fuel-based infrastructure, the feasibility of meeting 100% renewable energy goals for some cities will be challenging if not impossible. There are several reasons for this:

- Such goals require concerted efforts over long periods of time and must be continuously supported by multiple governmental administrations.
- Reaching the goals may require public–private partnerships with local corporations and energy suppliers. Some local governments are unfamiliar or uncomfortable with the process of creating and supporting such collaborations.
- While some cities may be able to meet the goal with little cost, most will find the goal costly.
- Existing electrical power infrastructure may not have been amortized. While renewable generation might be in some cases less expensive, changes over time could dramatically increase electrical rates.
- If costs for electricity increase substantially as a result of the changeover to renewables, ratepayers and taxpayers might challenge the goal periodically, creating political conflicts.
- Given the regional nature of renewable energy availability, creating and maintaining dispatchable electric baseload power using only renewable energy can be difficult. Providing a diverse and workable set of DERs many also prove challenging.
- The goals often exclude mobility fuels, such as diesel and gasoline, whose regulation may be beyond the reach of the local governmental entity.

Goals to transition to 100% renewables appear artificial and, in some cases, unrealistic. Transitioning away from fossil fuels completely may actually not be necessary in all instances to meet local environmental, carbon reduction, and economic goals. Should the targets be less? Perhaps, less challenging incremental goals targeting reductions of 40% to 80% in fossil fuel shares should be established for some communities as an alternative. Regardless, the institution of goals to incorporate greater amounts of renewable energy suggest that there will be a growing demand for community microgrids.

TYPES OF MICROGRIDS

Microgrids can be classified by their applications. Examples include mobile microgrids, utility distribution microgrids, community microgrids, military microgrids, industrial microgrids, campus microgrids, virtual power plants, and others.

Mobile Microgrids

A feature of many electric supply configurations is that their infrastructures are fixed in place and incapable of mobility. Once the grid infrastructure fails, they are unable to provide local power to a targeted group of loads. A *mobile microgrid* is defined as an independent, deployable power solution that can be used to supplement power sources in the event of a disruption to grid-supplied power [7]. The primary advantages of mobile microgrids include their smaller size, transportability, mobility, and adaptability. They can be readily placed into operation to provide local power. Many, such as fuel cell systems, are modular and can be deployed by truck or ship. If needed, several mobile microgrids can be linked together to create a larger power source [7]. These features make them useful in the event of power outages due to extreme weather events, for the provision of temporary power to secluded locations, or for special events or military operations.

Traditional types of generation for mobile microgrids include diesel or natural gas gensets, solar PV installations, or fuel cells. Special conditions create unique opportunities for mobile microgrids. A consideration for military field operations is that noise, exhaust, and the infrared signature from a mobile diesel generation system might reveal a stationary location to an enemy. A mobile microgrid with solar PV and battery storage might be preferable.

There are many configurations available for mobile microgrids. One manufacturer utilizes a solar hybrid inverter and charge controller system that can autonomously configure the DERs into a range of different variations to provide a resilient, efficient, clean, and cost-effective combination of energy resources at any given time [7]. Their advanced mobile microgrid incorporates a 9 kW solar photovoltaic energy, 144 kWh lithium-ion battery energy storage system, a 10 kW methanol reformer-based, hydrogen fuel cell system, plus an industrial ethernet switch for control and remote connectivity—all transportable in a 20′×8′ intermodal freight container [7].

Local Service Microgrids

Many communities view microgrids as a way to provide electricity for critical services during a crisis and enable emergency response (e.g., police, fire, and local hospitals). Microgrids are capable of providing power to local water supply and sewage systems. Local microgrids have multiple customers in close proximity with demand resources on the consumer side of delivery points. They can be designed to meet community objectives and may be consumer owned (e.g., by the locality). They are locally configured with a central distribution system, multiple generation resources and energy storage technologies to improve resiliency and reduce production costs (see Figure 9.1).

The value of local microgrids goes much deeper; not only may a microgrid deter economic calamity, but it may also enhance community prosperity [5]. Microgrids draw coveted high-tech industries to cities by offering the reliable, premium power and cleaner energy services that commercial enterprises require. In addition, many communities are seeking proven approaches to address climate adaptation and reduce greenhouse gas emissions, and are investing in more sustainable energy delivery systems [5].

Community and Local Microgrids

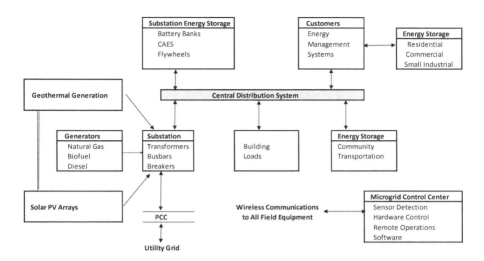

FIGURE 9.1 Configuration for community microgrids with energy storage.

MILITARY MICROGRIDS

Military microgrids have evolved from strategic decision-making and policy-making directives. Military bases are venues for management, training, housing, and important tactical initiative launching pads. The U.S. military is the world's largest single consumer of energy. It has developed two key policies to deal with climate change: 1) reducing its contribution to global warming by reducing its greenhouse gas emissions; and 2) maintaining capabilities to assist foreign (primarily allied) governments with coping with extreme disasters and to offer emergency assistance when directed [8]. Established policies include the 2012 and 2014 versions of the U.S. Department of Defense *Climate Change Adaption Roadmap*. It mandates more aggressive adaptation and sustainability measures requiring base commanders to integrate climate change projections into their installation master plans and institutes directives for "climate-specific plans and guidance" in procurement activities [8]. U.S. military bases are also under mandates to have the capability to maintain electrical power for mission-related services for as long as two weeks should external grid-supplied power be unavailable or disabled. Microgrids for the military can serve both base requirements and tactical efforts in the field.

Energy assets and systems for military operations are in a state of constant adaptation. Priorities for each support system can change in a moment, creating a need for electrical generation mobility. In many cases, the military places premiums on power accessibility, reliability, and equipment durability. Power systems can be disrupted despite being served by existing distribution grids. For these reasons, military microgrids have received renewed attention. They are being developed to support mission-critical facilities and are expanding at many military bases. Diesel generation systems, once considered a mission mainstay, are now considered by military logistics experts to be a weakness. Diesel fuel use has been proved to be highly questionable in field operations, especially when supply chain costs might include risking the lives of service personnel. Transporting fuel in war zones is one of the most dangerous jobs in the world and diesel fuel can cost up to $400 per gallon by the time it is delivered to front line units [9]. Substantial cost savings are possible when renewable energy can be forward deployed to replace diesel.

Key drivers for military microgrids include cost savings, reduced risks due to energy pricing variability, greater deployment of renewable energy, and improved energy security [10]. Driven in part by the doctrine of energy surety, military microgrids are characterized by diversity of mission, changing technologies, the need for just-in-time (JIT) integration, and the imperative to reallocate resources rapidly as missions vary and assets change [11]. Often bases may be located in or near urban centers, yet in times of grid failure, military operations must continue. Loss of power to mission-support operations present tactical risks to command response capabilities during national emergencies.

For military microgrid applications, there is a premium for energy systems that can reconfigure themselves within limits [11]. Military needs are best met through solutions that do not require constant reconfiguration and can be reused at multiple military installations [11]. They also prefer decentralized power systems with energy storage capabilities near living and working areas. This prevents the potential of the destruction of a single power plant causing complete loss of base power. Cybersecurity defenses to protect power generation facilities are more readily dispatched. These mission-critical capabilities provide strong arguments for microgrids to be considered as the first option for providing electricity.

Military bases are planning to integrate a variety of DERs into their microgrids. The Marine Corps Air Station Miramar in San Diego, California is incorporating new and existing resources that include a solar PV (1.6 MW), landfill gas (3.2 MW), diesel generation (6.5 MW), and a natural gas power plant [10]. A diagram of the Fort Sill, Oklahoma microgrid configuration for (20 kW solar, 2.5 kW wind power, 480 kW gas/diesel, 56 kW lithium-ion storage) is shown in Figure 9.2 [12]. Other U.S. examples of military microgrids can be found at Camp Pendleton South in California (525 kW solar, 200 kW gas/diesel), Fort Carson in Colorado (2,000 kW solar, 3,000 kW gas/diesel with integrated vehicles), Joint Base Pearl Harbor Hickman in Honolulu, Hawaii (146 kW solar, 50 kW wind power), Maxell Air Force Base in Alabama (1,300 kW gas/diesel), and West Point in New York (249 kW solar), among many others [9].

Industrial Microgrids

Today's industrial microgrids are often configured similarly to those industrial sites that produced their own power prior to the development of electrical grids. Cosidine et al. [11] describe industrial microgrids as focusing on process loads, costs, and energy resources:

> Industrial sites with high power requirements have long relied on site-based generation. For some, such as aluminum producers, electric power dwarfs all other supplies. If a site has power requirements similar to the capacity of commercial generating plants, in-sourcing this generation is a natural decision. Some processes, notably in

Community and Local Microgrids 147

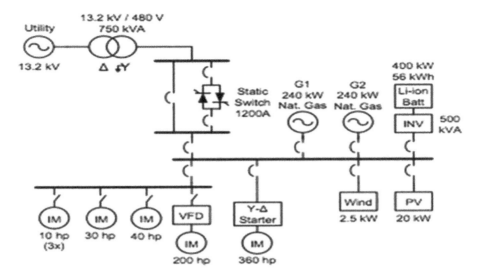

FIGURE 9.2 Microgrid for military base [12].

chemical processing and in regulated pharmaceutical environments, are subject to very large costs for power interruption. A small interruption in power may cause very large process costs, in lost product, in equipment degradation, in lost certification, and in high start-up costs afterward. Other sites use large amount of energy, but in a form other than of electrical power. In particular, some plants rely on thermal energy. These may form wood-based products or be laundry facilities. Steam or hot water may drive a significant part of their activities. Once they have a boiler in place, using excess capacity to generate electrical power is a natural afterthought [11].

A key advantage of industrial microgrids is the inherent capability of co-locating energy generation with existing electrical loads. Industrial sites are usually owned by a single controlling management structure which enables microgrids to be more rapidly installed.

Utility Distribution Microgrids

There are a growing number of utilities that have begun smart grid deployments and are now grappling with a new set of technical, regulatory, and financial challenges that indicate that the industry is undergoing evolutionary changes [13]. Identified trends include the integration of DERs and their associated assets, and changing business and regulatory approaches that require a far more sophisticated, reliable electric network which involves greater participation by customers and third parties in energy management and electric generation [13]. Some electric utilities are experimenting with the development of microgrids to support operations and improve services. Many are being developed in partnership with local communities and stakeholders. Utility distribution microgrids are identified by ownership and their ability to supply a central distribution network. They typically connect many unaffiliated end-users with multiple types of generation and energy storage [14]. These microgrids can be utility owned, utility managed, or co-managed by the utility and microgrid owner.

Microgrids managed or co-managed by public utilities provide an alternative to potential major capital investments. Rather than developing additional generation and transmission infrastructure to supply a remote local area, a grid-connected utility microgrid can reduce costs of service for the local utility [15]. For example, if a new substation or substation upgrade is required to meet increasing electric demand or resolve power quality issues, a local microgrid with on-site generation could meet the requirement without the investment associated with conventional solutions [15].

Campus Microgrids

Campus microgrids typically link a number of buildings owned by a single entity (i.e., university or corporate) on a single site. These buildings may use combined heat and power generators to supply the majority of their heat and electricity [14]. One of the best-known and most sophisticated microgrids in the U.S. is the 42 MW microgrid operated by the University of California at San Diego (UCSD), where administrators have formed public–private partnerships with federal agencies and utilities to incubate smarter, cleaner technologies [16]. The mission of this microgrid is to "promote multiple objectives, including lower energy costs, improved resiliency, and enhanced reliability" [16]. The university microgrid uses a diverse set of generation sources: a 2.8 MW fuel cell powered by waste methane from a local wastewater treatment plant, a 4 million gallon (15.1 million liter) thermal storage system, 2.0 MW of photovoltaics, plus 35 kW of solar concentrating photovoltaics [16]. The microgrid generates 92% of the university's power, and saves the campus $850,000 monthly in energy costs [16].

The microgrid for a college campus might be custom configured to restore power to select facilities that provide critical services in the event of a power outage. For example, it might be designed to restore power for emergency lighting, research labs, and security and medical facilities, but perhaps not the natatorium or academic offices [17]. With critical operations available, the college can maintain a minimal and crucial level of service so that when the central grid power is restored, the campus can more quickly resume normal operations [17].

The Alameda County Santa Rita Jail, about 70 km east of San Francisco, California, is an example of a campus microgrid with a peak demand of 3 MW (see Figure 9.3). The jail, designed to house 4,000 inmates, is the site of a microgrid demonstration project which has approximately 1.5 MW of solar PV, a 1.0 MW molten carbonate fuel cell, backup diesel generators that can function in grid-connected or island modes, and a 2 MW/4 MWh lithium-ion (LI) battery bank for load balancing [18]. The LI battery system and a sophisticated switch allow the jail to island and reconnect to the utility grid at will [18].

Virtual Power Plants

A *virtual power plant* (VPP) is a network of decentralized, medium-scale power generating units such as wind farms, solar parks, combined heat and power (CHP) systems, storage systems, and flexible power consumers [19]. By pooling diverse

Community and Local Microgrids 149

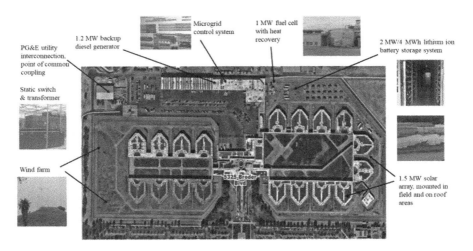

FIGURE 9.3 Configuration of the Alameda County Santa Rita Jail microgrid.

generation assets, and linking them through a central control center, the VPP behaves in relationship to the grid as a single controllable entity, similar to a microgrid. The interconnected units are remotely dispatched through a common control center and though interconnected are independent in their operation and ownership [19]. A VPP uses advanced software and technology to monitor and coordinate power generation assets and controllable loads, simulating the processes that would be expected from a conventional electrical power plant. The bidirectional data exchange between the individual plants and the VPP not only enables the transmission of control commands but also provides real-time data on the capacity utilization of the networked units [19]. During periods of peak load, the VPP relieves the load on the grid by smartly distributing the generated electricity [19]. The power generation and consumption of the networked units in the VPP can be traded on an energy exchange [19]. In some configurations, it is possible to sell and purchase units of electricity using a smartphone. Microgrids can be configured to operate either as a VPP or to mimic their capabilities and services. Similar to conventional power plants, cyberattacks are a threat to VPPs.

The real-world working applications for virtual power plants are by definition diverse. In the U.S., a VPP may not necessarily involve the actual deployment of electrical generation sources. VPPs might access utility demand response and peak pricing programs as primary resources. When aggregated, the VPP mimics the characteristics of a traditional power plant delivering peak capacity, electrical energy, or grid reliability regulation services as needed by a utility or independent system operator (ISO) [20]. In Europe, a VPP typically refers to aggregating supply side resources, most often a diverse pool of renewable distributed energy generation (RDEG) and/or wholesale renewable energy sources [20]. More confounding, the term VPP might also refer to the ability of commercial consumers in countries such as Denmark to purchase capacity at the wholesale level via an auction from baseload fossil fuel facilities for short periods of time. Statkraft, a Norwegian company, has operated one of Europe's biggest power generation facilities as a VPP since 2011 [21]. Their VPP in Germany has a 12,000 MW capacity, enough power for five million homes [21].

VPPs often provide services to the host grid in exchange for preferred pricing. Flexible power-consuming equipment (e.g., motors and pumps) can be operated on optimized price schedules by using electricity when it is inexpensive and electric demand is low [19]. The VPP central control system helps stabilize grid power before balancing is required [19]. If a grid imbalance is imminent, the signals from the system operators are processed in the central control system and converted into control instructions for the pre-qualified units effectively keeping the grid in balance by delivering frequency control reserves [19]. If an unexpectedly high feed-in develops, it is possible to shut down generation assets within seconds and avert critical grid situations [19]. When excess supply is available it can be stored. According to Andreas Bader, vice president for sales for Statkraft, "We can connect batteries from Spain with wind farms in Germany, and that makes it scalable" [21].

EXAMPLES OF COMMUNITY AND LOCAL MICROGRIDS

Traditionally, microgrids have served only one user, such as a university or hospital, and provide the host campus with an independent energy supply. In some U.S. states (e.g., New York), the community-based focus is to connect multiple users through DERs with more reliable and secure energy sources [22]. Community microgrids build upon existing electrical system infrastructure and equipment, connect multiple users in a neighborhood, and provide local power generation and distribution [22]. The process of building a community microgrid typically involves beginning with a basic configuration that addresses the community's key priorities or pain points [23]. These are identified problems or issues that have come to the attention of the community and require resolution. To address these, communities often begin by constructing a *foundational microgrid* [23]. Incremental improvements and expansions of the foundational microgrid evolve over time ultimately forming a more complex microgrid [23]. Examples of operational community microgrids include those on Kodiak Island, in Borrego Springs, and on Long Island which are next discussed.

KODIAK ISLAND

Kodiak Island in Alaska is often used as an example of a remote community-based island microgrid. It is one of the largest such microgrids in the U.S. to use renewable energy. Its microgrid operates using 5 MW hydroelectric power, a large 9 MW windfarm, a gas/diesel gen-set system, and 5 MW of energy storage. The system is rated at 75 MW and provides power to the island's 15,000 residents using 99% renewable energy. Storage systems have lead-acid batteries and flywheels. Two flywheels are used stabilize the microgrid and a highly dynamic industrial load [15]. These flywheels: 1) provide frequency regulation and demand smoothing; 2) relieve stress on existing battery systems, extending their useful life; 3) manage intermittency from the wind turbine generators; and 4) reduce reliance on diesel generators [15].

The use of renewables combined with storage technologies caused electrical costs to decline to about $0.14/kW, saving the community roughly $4 million annually compared to using diesel-fueled generation systems [24].

BORREGO SPRINGS MICROGRID

Borrego Springs, California is somewhat remote and supplied with electricity by a single transmission line. If the electrical feed de-energizes, the community has no electricity. To prevent loss of power, the community developed the first microgrid in the U.S. to rely only on solar energy and storage during planned outages [14]. Many of the microgrid's DERs are owned by the residents.

The Borrego Springs Microgrid, serving 2,780 customers, was developed in an arid California location subject to dry conditions and extreme temperatures. Average maximum/minimum winter temperatures are 69°/43.4°F (20.6°/6.3°C) and 106.8°/74.9°F (41.6°/23.8°C) in July [26]. This solar-storage microgrid is a proof-of-concept model of an unbundled utility microgrid. Distribution assets are owned by San Diego Gas and Electric (SDG&E), a public utility, while some of the distributed power generation assets are owned by customers [26]. Thus, it provides an example of an *unbundled utility microgrid*. As it provides services to the community, the Borrego Springs Microgrid is also considered to be a *distribution company microgrid*.

The total microgrid installed capacity is about 4 MW (see Figure 9.4), with the main technologies being two 1.8 MW diesel generators, a large 500 kW/1,500 kWh battery at the substation, three 50 kWh batteries, six 4 kW/8 kWh home energy storage units, about 700 kW of rooftop solar PV, and 125 residential home area network systems [27]. The microgrid control systems incorporate supervisory control and data acquisition (SCADA) capabilities on all circuit breakers and capacitor banks, feeder automation system technologies (FAST), and outage management systems which provide price-driven load management for its customers [27]. The cost of the project was $15.2 million, supported by the U.S. DoE ($2.8 million), San Diego Gas and Electric ($7.5 million), and public–private sources ($4.9 million) [28].

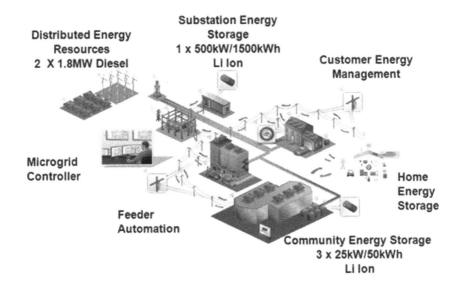

FIGURE 9.4 Borrego Springs microgrid. (Source: National Energy Technology Laboratory [25].)

LONG ISLAND COMMUNITY

The Long Island Community Microgrid Project (LICMP) is being developed as a multi-stage project in East Hampton, New York. It is a community that experiences recurring electrical outages primarily due to storm damage. A solar PV siting survey was performed and found over 30 possible sites, such as vacant land, rooftops, and parking areas, with a total potential capacity of 32 MW. The survey led to a community microgrid design that would use local solar and energy storage for electrical generation.

Key partners in the LICMP are PSEG Long Island, Long Island Power Authority, Suffolk County Water Authority, and the Springs Fire District [29]. The community microgrid benefits over 3,300 account holders and a population of roughly 10,000 local residents. The project goals include: 1) reducing dependence on the transmission grid and oil-based generators; 2) increasing the penetration of local renewable energy; 3) maintaining electric services for critical loads during grid outages; 4) demonstrating the feasibility of using energy storage in utility grid operations to increase local renewable generation; and 5) decreasing fossil fuel consumption and transmission costs [29]. The Long Island Community Microgrid includes the following features [30]:

- 15 MW of new solar PV generation
- 5 MW/25 MWh energy storage facility plus three smaller energy storage facilities
- enhancing monitoring, communication and control systems
- provision of indefinite renewables-driven backup power for multiple critical loads that include a fire station and water pumping stations
- providing utility-scale peak shaving
- demonstrating robust community microgrid capabilities over a substation grid area
- showcasing how to design and operate electric community microgrids with local renewables and other DERs [30]

The benefits provided by of the LICMP include [30]:

- providing local renewable energy with reduced dependence on centralized, non-renewable power, and oil-fueled peak generation facilities
- reducing peak power demand by over 8 MW
- avoiding over $38 million in local transmission systems upgrade costs with the potential of avoiding an additional $300 million in future transmission upgrades if the LICMP is replicated within the region
- reducing capacity charges and energy arbitrage costs
- generating $32 million in wages and other economic value during the construction phase of the project, with millions more during ongoing operations [30]

A catalyst for this project was the impact of superstorm Sandy in 2012, which wiped out parts of New Jersey and New York [31]. The damage to utility infrastructure prompted the community to create the microgrid and ultimately provide half of its grid-area electric power requirements from solar power [31].

SUMMARY

Community microgrids are miniature versions of the larger grid to which most are interconnected. According to the Clean Coalition, an organization that supports microgrid initiatives, a community microgrid is a "coordinated local grid area served by one or more distribution substations and supported by high penetrations of local renewables and other DERs, such as energy storage and demand response" [32]. Community microgrids offer a flexible and scalable means of expanding the benefits of local distributed energy solutions to provide resiliency, reduced carbon emissions, and energy cost management within networks of energy users [22].

Microgrids are commonly categorized by the types of local areas and communities they serve. Examples of community-scale microgrids include mobile, local service, utility distribution, military, industrial, and campus microgrids plus virtual power plants. VPPs are typically a network of decentralized, medium-scale power generating units that link renewable and fossil-fired generation with flexible power consumers using a central control center. Community microgrids typically serve an entire substation, are installed on the consumer side of the meter, provide backup power, and are more readily replicated and scalable across the distribution grid when compared to traditional microgrids [32].

Motivations to develop community microgrids vary widely. Communities and cities are seeking ways to act locally to solve large-scale problems. To be viewed as meaningful, this requires finding ways to respond to problems associated with energy and environmental issues. Other communities seek improved reliability and lower costs. Community microgrids are seen as solutions that have the potential to supply electrical power in periods when the central grid fails. They also reduce utility costs, provide new infrastructure and employment, and expand the local government's tax base [33]. Many communities see local microgrids as a means of meeting their goals to reduce both greenhouse gas emissions and fossil fuel consumption. This is accomplished by relying on renewables for electrical generation coupled with the use of sophisticated control systems and energy storage technologies. Regardless, it is common that community microgrids deploy both renewables and fossil fuel (notably natural gas and/or diesel) in their operating configurations to generate electricity.

Military microgrids are motivated by the need to provide mission-critical field support, maintain base readiness, or meet policy goals. Dependency on grid-based electricity is problematic for military operations in an era when central generation and transmission can be easily targeted. Climate change is also a driver for military bases. U.S. military officials view climate change as a "secondary but insidious threat, capable of aggravating foreign conflicts, provoking regional instability, endangering American communities, and impairing the military's own response capabilities" [8]. Renewable energy microgrids help military bases reduce their reliance on diesel fuel and reduce their carbon footprint.

There are many examples of community microgrids. Successful examples of microgrid deployment include those for the communities of Kodiak Island, Bonita Springs, and Long Island. Most are fueled by using combinations of renewables and fossil fuels. The Isle of Eigg off the coast of Scotland has its own microgrid that uses diesel generators, 110 kW of hydropower with two generators, 2 kW provided by four wind turbines, plus 32kW of solar PV [34]. The system offers improved load management and continuous electricity with 95% from renewable energy sources [34].

Successful microgrid deployment is achieved with community involvement and support. Microgrids can be designed to be locally responsive and configured to meet local needs. They can more easily fulfill the design purposes of local communities than by grid integration based on central generation and far-off service centers [11]. Microgrids offer more options for balancing intermittent energy generation than do larger grids, because the tradeoffs are visible and local [11].

REFERENCES

1. Lantero, A. (2014, June 17). How microgrids work. https://www.energy.gov/articles/how-microgrids-work, accessed 21 December 2019.
2. Microgrid Institute (2014). About microgrids. http://www.microgridinstitute.org/about-microgrids.html, accessed 11 November 2018.
3. Clean Coalition. Community microgrids. https://clean-coalition.org/community-microgrids, accessed 21 December 2019.
4. International Renewable Energy Agency (2015). Off-grid renewable energy systems: status and methodological issues, working paper. https://www.irena.org/DocumentDownloads/Publications/IRENA_Off-grid_Renewable_Systems_WP_2015.pdf, accessed 18 February 2020.
5. Microgrid Knowledge and the International District Energy Association (2015). Community microgrids, a guide for mayors and city leaders seeking clean, reliable and locally controlled energy. https://microgridknowledge.com/white-paper/microgrid-policy-guide, accessed 24 January 2020.
6. Sierra Club (2018). 100% commitments in cities, counties and states. https://www.sierraclub.org/ready-for-100/commitments, accessed 21 December 2019.
7. Business Wire. Faith technologies and Excellerate Manufacturing launch mobile microgrid energy solution. https://www.businesswire.com/news/home/20180723005664/en/Faith-Technologies-Excellerate-Manufacturing-Launch-Mobile-Microgrid, accessed 27 November 2019.
8. Klare, M. (2019). *All Hell Breaking Loose*. Metropolitan Books: New York.
9. Microgrid Projects. U.S. military microgrids, why? http://microgridprojects.com/military-microgrid-army-navy-air-force-microgrids-drivers, accessed 20 December 2019.
10. Black and Veatch (2017, June 13). Top 3 benefits of military microgrids. https://3blmedia.com/News/Top-3-Benefits-Military-Microgrids, accessed 3 December 2018.
11. Cosidine, T., Cox, W., and Cazalet, T. (2012). Understanding microgrids as the essential architecture of smart energy. https://www.gridwiseac.org/pdfs/forum_papers12/considine_paper_gi12.pdf, accessed 20 December 2019.
12. Microgrid Projects. U.S. military microgrids – why? http://microgridprojects.com/military-microgrid-army-navy-air-force-microgrids-drivers, accessed 22 December 2019.
13. U.S. Department of Energy (2014, August). 2014 smart grid system report. https://www.smartgrid.gov/files/2014-Smart-Grid-System-Report.pdf, accessed 29 November 2019.

14. Gimley, M. and Farrell, M. (2016). Mighty microgrids. Institute for Local Self-Reliance. https://ilsr.org/wp-content/uploads/downloads/2016/03/Report-Mighty-Microgrids-PDF-3_3_16.pdf, accessed 23 December 2019.
15. GTM Research (2016, February). Integrating high levels of renewables into microgrids – opportunities, challenges and strategies, white paper. http://www.sustainablepowersystems.com/wp-content/uploads/2016/03/GTM-Whitepaper-Integrating-High-Levels-of-Renewables-into-Microgrids.pdf, accessed 24 January 2020.
16. Cohen, R. (2013, November 5). State and local energy report. http://stateenergyreport.com/2013/11/05/in-the-quest-for-a-more-resilient-grid-microgrids-offer-solutions, accessed 28 November 2018.
17. Wood, E. (2018, November 4). Microgrid benefits: eight ways a microgrid will improve your operation… and the world. https://microgridknowledge.com/microgrid-benefits-eight, accessed 23 December 2019.
18. Berkeley Lab (2018). Microgrids at Berkeley Lab, Santa Rita Jail. https://building-microgrid.lbl.gov/santa-rita-jail, accessed 23 December 2019.
19. Next. Virtual power plant. https://www.next-kraftwerke.com/vpp/virtual-power-plant, accessed 28 November 2019.
20. Kashyap, S. (2017, October 12). What is the difference between microgrids and virtual power plants? https://www.quora.com/What-is-the-difference-between-microgrids-and-virtual-power-plants, accessed 11 November 2018.
21. Ziady, H. (2019, November 7). No wind? No sun? This power plant solves renewable energy's biggest problem. CNN Business. https://www.cnn.com/2019/11/07/business/statkraft-virtual-power-plant/index.html, accessed 23 December 2019.
22. New York State (2018). Microgrids 101. https://www.nyserda.ny.gov/All-Programs/Programs/NY-Prize/Resources-for-applicants/Microgrids-101, accessed 22 December 2019.
23. Agate, W. (2019, November). Community microgrids, best practices and the promising future outlook of localized energy. *Distributed Energy*, pages 13–15.
24. Grimley, M. and Farrell, J. (2016, March). ILSR's energy democracy initiative. https://ilsr.org/wp-content/uploads/downloads/2016/03/Report-Mighty-Microgrids-PDF-3_3_16.pdf, accessed 26 January 2020.
25. Burger, A. (2016, December 25). Borrego Springs microgrid: from proof of concept to project development showcase. http://microgridmedia.com/borrego-springs-microgrid-proof-concept-project-development-showcase, accessed 30 November 2019.
26. Burger, A. (2017, June 23). Renewable, hybrid island microgrid projects under way in Aruba, U.S. Virgin Islands. http://microgridmedia.com/renewable-hybrid-island-microgrid-projects-way-aruba-us-virgin-islands/#top_ankor, accessed 3 December 2019.
27. U.S. Department of Energy (2019). Microgrids at Berkeley Lab, Borrego Springs. https://building-microgrid.lbl.gov/borrego-springs, accessed 26 November 2019.
28. KEMA (2014, February 3). Microgrids – benefits, models, barriers and suggested policy initiative for the Commonwealth of Massachusetts. Burlington, Massachusetts. http://nyssmartgrid.com/wp-content/uploads/Microgrids-Benefits-Models-Barriers-and-Suggested-Policy-Initiatives-for-the-Commonwealth-of-Massachusetts.pdf, accessed 29 May 2020.
29. Microgrid Projects. Long Island community microgrid project in East Hampton. http://microgridprojects.com/microgrid/long-island-community-microgrid-project-in-east-hampton, accessed 26 November 2019.
30. Valentine, J. (2016, March 29). Long Island community microgrid project completes stage 1 of the NY Prize. https://clean-coalition.org/news/long-island-community-microgrid-completes-stage-1-ny-prize, accessed 26 November 2019.
31. Silverstein, K. (2019, March 7). Montecito community microgrid emerges from mudslides and wildfire. https://microgridknowledge.com/community-microgrid-fire, accessed 27 December 2019.

32. Clean Coalition. Community microgrids compared to traditional microgrids. https://clean-coalition.org/community-microgrids, accessed 30 November 2019.
33. Microgrid Knowledge (2019). Community microgrids, a guide for mayors and city leaders seeking clean, reliable and locally controlled energy. https://microgridknowledge.com/.../guide-to-community-microgrids, accessed 13 January 2019.
34. U.S. Department of Energy (2019). Microgrids at Berkeley Lab, Isle of Eigg. https://building-microgrid.lbl.gov/isle-eigg, accessed 26 November 2019.

10 Ghana's Transition to Renewable Energy Microgrids

Ishmael Ackah, Eric Banye, Dramani Bukari, Eric Kyem, and Shafic Suleman

INTRODUCTION

Microgrids are considered to be one of the best alternatives for providing electricity to remote locations with small populations. They can be designed to use local resources, scaled to the required loads, and rapidly deployed. Using conventional electricity transmission networks as a means of promoting access to electricity faces geographic, financial, and technical challenges. This is because many rural communities are sparsely populated and located far from grid connections. Such challenges have led to increased interest in off-grid and microgrid solutions. Ghana is not an exception to this trend. The country established a target of achieving universal access to electricity by 2020. The government and private sector companies invested in off-grid and microgrid projects to provide services to the remaining areas and island communities. There are notable differences in the services and results of the government model compared to the private model of ownership and operation.

This chapter highlights the viability of solar photovoltaic (PV) microgrids for the rural electrification in Ghana. It offers an analysis of the regulatory and fiscal situation and offers recommendations for the use of renewables within a microgrid regulatory framework. To these ends, this chapter examines the microgrids deployed by both the government and private investors. It assesses the differences in tariff structures, customer services, and reliability. Building upon the work of past researchers, this analysis highlights the viability of microgrid solar PV systems for rural electrification systems. The chapter reports on the findings from a study that uses a purposeful sample of households and applies the subject evaluation method for an overall assessment. These findings indicate that expenditures on electricity using microgrids were lower compared to the alternative of using kerosene and dry cell batteries.

ACCESS TO ELECTRICITY IN AFRICA

Universal access to affordable and reliable electricity in the least developed countries is a top priority in meeting the goals of sustainable development. About one billion people globally live without electricity, more lack sufficient or reliable access to electrification, and about three billion cook or heat their homes with polluting fuels.

Renewable energy (RE) is one of the vital ways to help countries develop modern, low cost, secure energy systems. Solar PV microgrids are a viable method to close the electricity access gaps in southern Asia and sub-Saharan countries [1,2].

Solar generation technologies offer a rapid, cost-effective means of providing electricity to 600 million Africans who presently lack access to modern utility-scale, grid-provided electricity [3]. Solar PV offers a modular solution for both on- and off-grid applications which can provide electricity to single homes or microgrids for loads ranging from several kilowatts to many megawatts [3].

Isolated microgrids utilizing solar PV can provide electricity to entire communities in a single project, enable interconnections with the local grid at a later date, plus reduce operational costs [3]. Africa's microgrids are usually small in scale and range from 8 kW to 10 MW. However, large-scale industrial customers such as industrial operations offer opportunities for larger solar PV systems up to 40 MW. Solar PV microgrids offer important economic benefits for rural communities as either the sole source of electricity generation or in hybrid configurations with other generation sources. In Africa, installation costs for standalone solar PV microgrids or solar PV-hybrid microgrids range from $1.9/W (U.S.) to $5.9/W for systems greater than 200 kW [3].

Ghana's Renewable Energy Act

Ghana enacted its renewable energy act (Act 832) in 2011 to promote the development, utilization, management, and supply of heat and power delivered in an efficient and environmentally sustainable manner. The Act seeks to improve energy access from Ghana's RE sources. The country's access rate in 2018 was 85%, which means about 4.5 million people lack access to electricity. It supports the development of a framework for the utilization of RE and provides enabling legislation to attract RE investments.

Act 832 enlightens two important issues that require investigation: 1) how to attract renewable energy investments to Ghana; and 2) how to improve access to renewable energy. The focus of our study was to provide a viability analysis for RE projects to clarify the challenges and risks for investors wanting to understand the RE investment environment in Ghana. The study focused on solar PV microgrid investment viability in select rural communities that lack access to Ghana's national grid. This will inevitably improve energy access in those rural communities which lack grid-supplied electricity.

MICROGRID SOLAR PV SYSTEM VIABILITY IN GHANA

In Ghana, more than 85% of the communities with populations greater than 500 have access to grid-supplied electricity [4]. About 15% of Ghana's population live in rural and sparsely populated communities and are not connected to the national grid [5]. Many communities with populations under 500, such as those in the Volta Lake area and in isolated lakeside communities, lack access to electricity. Grid-based connections to remote communities require costly underwater cables, transmission lines, and distribution lines. Dispersion and maintenance issues make grid connections uneconomical in such instances. Therefore, the use of distributed generation (DG) is often the most technically and economically viable solution [5].

Ghana's central government is targeting the construction of 55 renewable energy microgrids by 2020 with the ultimate aim of developing 300 microgrids by 2030. The targeted locations for microgrid deployment are lakeside, island communities and rural off-grid communities [5]. In 2016, five microgrid projects in island communities using solar and solar–wind hybrid technologies with diesel gensets were deployed.

One theme of the fourth annual Renewable Energy Fair organized by the Energy Commission of Ghana in October 2018 was how to exploit energy resources at the district level. One of the technical sessions focused on whether the state or the private sector is best positioned to use RE to deliver access to electricity. The managing director of the local microgrid company indicated that the main challenges to microgrid development in Ghana are the viability of a project's finances and regulatory support. Act 832's aim of attracting and promoting private sector investments to supply RE to rural communities must focus on providing a viable business environment for private sector participation to meet the country's energy requirements.

The Energy Sector Management Assistance Programs (ESMAP) developed five business models to generate and distribute electricity to these remaining communities. Advantages and disadvantages for the investment models are shown in Table 10.1 [4]. None emerges as a universally superior model. The ability to meet the requirements of Act 832 is hampered by the ESMAP (2016). The key disadvantage of the private ownership model is a lack of investment because of revenue collection risks, high transaction costs, and concerns regarding the technical and managerial capacities of community-based models. This highlights the main problem of our study: how viable are solar PV microgrids for meeting Ghana's rural community electricity requirements?

PERSPECTIVES ON MICROGRIDS

Solar PV systems are nascent infrastructure projects in Africa. They often face problems and business development challenges. Solar microgrid projects require the engagement of many diverse stakeholders. Large projects have higher costs making them less attractive as investment opportunities. The lack of development cost information is a problem as there is no coordinated effort to document the installed costs of solar PV projects in Africa or across specific market segments. Gollwitzer et al. relied on an analytical framework and the common-pool resource theory to analyze rural microgrids [6]. They considered enabling sociotechnical and political conditions, resource system characteristics (governance of local resources by, e.g., elite groups, donors, or international non-governmental organizations, governments, and civil groups), institutional arrangements, and environmental conditions that influence the viability of rural microgrid projects [6]. Rule-of-use issues surfaced which prohibit using energy intensive appliances (e.g., irons) or installing technologies to control usage. Time-of-use was identified as being an important consideration as businesses were encouraged to operate during the daylight hours when household electric demand was lower and discouraged from operating during evenings when

TABLE 10.1
Models for Microgrids Investments [4]

Microgrid business model	Description	Advantages	Disadvantage
Public model	Public sector provides generation and distribution.	Customers have low tariffs.	No role for private sector; highly reliant on cross-subsidies.
Private model	Private sector provides generation and distribution.	Less reliant on cross-subsidies.	High revenue risks, high transaction cost, less interest from the private sector; likely to result in high cost reflective tariffs.
Mixed model (public generation, public distribution, private management model)			Possible conflicts over the long-term regarding responsibility on reinvestments; lack of precedents.
Power purchase agreement (PPA) model (generation private (on the basis of a PPA), public distribution model		Clear division of responsibilities; customers have lower tariffs.	Require recurrent subsidies (can be through cross-subsidies).
Community model	Community acceptance		Concerns regarding technical and managerial capacities.

more electricity is required by households (i.e., for lighting and entertainment). There were concerns about tariff structures and the variable willingness and ability of customers to pay for services. Their study failed to capture an in-depth analysis of participatory governance and energy access in rural microgrid electrification.

The International Renewable Energy Agency (IREA) identified three issues that needed to be addressed to resolve data collection problems in Africa [3]. First, a standardized data collection system was needed that was agreeable to and implemented by stakeholders. Second, governments, international development organizations, regulatory authorities, development banks, lending institutions, and others that provide public financing and financial assistance to RE projects must be included in the data collection process. Third, coordinated support for RE projects to include sharing information on costs and trends needs to be encouraged [3].

Unlike other small-scale electrification approaches such as pico-lighting or solar home systems (SHS), microgrids have greater potential to enhance local economic growth and productivity [7]. For remote locations, solar PV microgrids combined with battery storage are often cost-competitive compared with diesel-based power generation. Alternatives to the solar PV microgrids include SHS Table 10.2.

In India, microgrids offer opportunities for rural electrification yet there has been little private investment [7]. Microgrids provide a collective solution at a relatively

TABLE 10.2
Theoretical Review of Global and African Perspectives

Author	Title	Methodology	Findings/recommendations
International Renewable Energy Agency (2016) [3].	Solar PV in Africa: costs and markets.	Sidewise analysis.	*Findings*: Solar PV microgrids are an attractive rural electrification option. Solar PV project cost data are not systematically collected in Africa. *Recommendation*: Coordinate efforts to collect the installed costs of solar PV, across markets in Africa to improve policy-making and share experiences among countries and regions.
Comello et al. (2016) [7].	Enabling microgrid development in rural India.	Lifecycle cost analysis and stakeholder interviews.	*Findings*: Microgrids offer the potential for rural electrification while benefiting economic growth and productivity. In India, developers do not need a license to operate a microgrid solar system. There are no legal or regulatory frameworks that specify what happens if a central grid is extended to an area covered by a microgrid. Microgrid investments are disadvantaged in central grid extensions because of subsidized central grid tariffs. *Recommendations*: Develop state-level regulations for microgrids that address these problems.
Bhattachayya S. (2014) [9].	Viability of off-grid electricity supply using rice husk: a case study of south Asia.	Financial analysis—levelized cost of electricity.	*Findings*: Regulatory uncertainties and the potential of grid extension can hinder the business potential of a microgrid power system. *Recommendations*: Expand microgrids and extend services to consumers who are willing to pay to increase the economics of scale and scope. The issues of regulatory uncertainties for microgrids should be solved to reduce the risk of access to project funds.

(Continued)

TABLE 10.2 (CONTINUED)
Theoretical Review of Global and African Perspectives

Author	Title	Methodology	Findings/recommendations
Bhattachayya C. and Palit, D. (2016) [8].	Microgrid based off-grid electrification to enhance electricity access in developing countries.	Analytical framework: structured interviews, focus group discussions.	*Findings*: Microgrid and off-grid systems are required to enhance universal access to electricity. The private sector must be involved. *Recommendations*: Provide regulatory clarification, strong institutions at the meso and local level with appropriate institutional arrangements, access to soft funding, and bundling of projects. Employ standard processes, metrics, and delivery of electricity to enable win–win situations for all key stakeholders.
Gollwitzer et al. (2018) [6].	Rethinking sustainability and the institutional governance of electricity access with microgrids: electricity as a common pool resource.	Semi-structured interviews.	*Findings*: No model of sustainable management of rural microgrids exists which contributes to high failure rates.
Knuckles, J. (2016) [12].	Business models for microgrid electricity in pyramid markets.	Business model analytical framework and interviews.	*Findings*: The study identifies 29 configurations of elements within microgrid business models that supply the large and growing unmet demand for electricity in these markets.
Arranz-Piera et al. (2018) [5].	Microgrid electricity service based on local agricultural residues: feasibility study in rural Ghana.	Field survey, interviews.	*Findings*: Residential electricity loads dominate rural electricity demand in Ghana. It is not profitable from the prospective of an entrepreneur with 100% private funding; however, it is profitable with a 35% subsidy. *Recommendations*: Develop solar PV to supplement biomass generation.

lower cost by offering basic electrical services. They increase local productivity and promote economic development [8].

Using a comprehensive lifecycle cost analysis Comello et al. determined that microgrids using solar PV are more economical than incumbent energy services available to households that lack central grid connections [7]. Entrepreneurs in India do not require certification or licenses to start a microgrid project. Investments in microgrids would be jeopardized if a central grid were extended because governmental electricity subsidies typically fail to cover the full costs of grid-supplied power. Therefore, in India threats of future central grid extensions create barriers to microgrid project financing and development [7,9].

Financing microgrid projects is challenging due to the limited number of bankable contracts with consumers. Governments and international agencies can support project financing through grants, loans, guarantees, and tax credits [9]. Regulatory supervision of microgrid projects is undefined in many south Asian countries. This creates potential for monopolistic exploitation, supply quality issues, and disputes between suppliers and consumers [9]. There is no legal or regulatory framework for rural microgrid projects that resolves what happens when a central grid is extended to areas already served by microgrids thus stranding the investments in off-grid assets [7,9,10]. Bhattachayya and Palit identified regulatory uncertainties and a policy vacuum that hinders the growth of solar PV microgrids serving rural areas [8].

Strengthening the regulatory environment is one of the prerequisites for attracting private investments in solar microgrid projects [8,9]. Microgrid electrification involves a natural monopoly, control schemes to protect consumers and investors, service quality monitoring, plus information incident reporting [8]. To be successful regulations must: 1) avoid confusion about the service areas; 2) protect investors from the threat of central grid extension; 3) ensure quality and reliability of services; 4) promote health and safety; 5) ensure transparency and flow of relevant information; and 6) ensure financial sustainability through tariffs and support systems.

Energy Requirements of Rural Communities

There is a lack of modern energy services in rural communities in sub-Saharan Africa where over 80% of rural people live without electrical power [10]. There is a need to accelerate access to electricity in remote and sparsely populated rural communities. Microgrids offer a modular and competitive solution [10]. Bhattacharyya and Palit noted that microgrids are a recent development for providing electricity to rural communities [8]. The challenges they face include business risks associated with unknown consumer markets, weak institutional arrangements, non-supportive regulatory and policy frameworks, limited access to low-cost financing, and inadequate local skills and capacities [8].

Solar PV microgrids can offer electricity to rural communities using three business models: lighting only, lighting plus, and anchor loads. *Lighting only* models use micro systems to provide lighting to rural households, small villages, hamlets, or concentrated populations. Each user gets two light points plus a mobile charging point with electrical demand limited to 10W per household. These solar energy applications are adequate to displace the harmful effects of using kerosene lamps.

The *lighting plus* model provides electricity to other households and the local community. Households with higher electricity demand and greater ability to pay for services can be supplied power while meeting commercial electrical demand. The *anchor load* model is a variation of the lighting plus option. It has an anchor electricity user (e.g., a telecommunication tower or local industry) that provides the base electric load and the excess supply is distributed to the community for their lighting systems [8].

Arranz-Piera et al. identified four main categories of electricity demand requirements in rural communities in a microgrid configuration: residential, institutional, commercial, and industrial [5]. Residential communities include private households that use electricity for lighting, air conditioning, refrigeration, and entertainment. Institutional consumption includes the electrical consumption of public institutions in the community [5]. Public lifting, water pumping, and electricity use in religious buildings, schools, and health centers are examples. Commercial consumption represents electricity consumed by small commercial entities such as dressmakers, shops, bars, and hairdressers. Their electrical consumption is related to each community's characteristics. Industrial consumption represents the electricity needed by small industrial consumers and their usage depends on their loads and operating cycles. Examples include water purification operations, irrigation farming, and food processing operations (cocoa, millet, shea butter), among others.

Gollwitzer et al. in their study assessed the social and political impacts of rural microgrid development in East Africa. They determined that electricity consumers in rural communities struggle financially and have less ability to pay tariffs that reflect the total cost of providing electricity [6]. For solar PV microgrid users, meter installation is expensive which means small DG systems must charge flat tariffs which creates incentives for increased usage by customers [6]. They also engaged the common pool resources theory which notes that social institutions can be formed to achieve sustainable management of rural microgrid systems [6].

Building upon the work of these past researchers, our analysis highlights the viability of microgrid solar PV systems for rural electrification in Ghana. We assess the regulatory and fiscal situation with a goal of recommending a RE and microgrid regulatory framework. In the absence of electricity supplied by a central grid, people in rural areas must rely on kerosene, diesel, or small RE generation. As an alternative, how viable are solar PV microgrids for providing electricity for isolated rural communities?

METHODOLOGY

Our study applied a cross-sectional method to examine the socioeconomic impacts of microgrids to provide off-grid electricity services in selected communities in Ghana. Using the renewable energy technology (RET) penetration barriers research methodology proposed by Painuly, interviews were conducted with consumers in selected communities, power sector representatives, and members of the Ghana Energy Commission [11]. A combination of structured questionnaires and interviews were used to study 300 respondents in eight communities in four regions in Ghana. The subjective evaluation method (SEM) was used to examine the impacts of access to microgrids on survey respondents. In addition, a descriptive analysis of the respondent's satisfaction levels and factors that inhibit or promote their productive use of electricity was conducted. The information collected from the field investigation was then vetted and captured using a CSPro template. The captured data were cleaned to avoid errors before analysis.

DISCUSSION OF FINDINGS

The categories of findings regarding microgrids in Ghana included policy and regulatory aspects, differences between private and government microgrids, plus differences in reliability, capacities, technologies, and tariffs. Social impacts included the various impacts of the projects on household expenditures, the work of women, and education.

MICROGRID POLICY AND REGULATION IN GHANA

Countries that have successfully developed local microgrids have carefully fostered the social, environmental, and economic conditions, needed to create a sustainable regulatory and financial scheme specific to the country. Ghana's central government through its Energy Commission has drafted microgrid regulatory policies which promote development in off-grid communities and regulate private sector engagement. These regulations apply to the development and operations of microgrids with generation capacities up to 1 MW. Persons and companies installing and operating microgrids that provide between 100 kW to 1 MW of DG capacity are required to obtain a license. Applicants are obliged to apply for and obtain an additional license if an existing system is expanded beyond 100 kW capacity.

PRIVATE AND GOVERNMENT MICROGRID SYSTEMS

Microgrid developments are relatively new in Ghana as most are less than five years old. Efforts to develop and increase access to electricity using microgrids in mostly rural communities face practical difficulties including system management, weather, and technical challenges. The ability of consumers to pay for the services provided is an economic concern. Successful operational models include private and government-managed systems.

Government-managed systems in Ghana currently operate in a few island communities as it would be uneconomical to connect them to the national grid. Microgrids in these communities offer the most viable option for providing electricity as they are isolated by Volta Lake. Four communities in four districts benefit from the government-managed microgrids: Pediatorkorpye in Ada East, Atigagorme in the Sene district, Kudorkope in Krachi East, and Agilakorpe in Krachi West. These projects began operating in 2016. Energy Commission data shows that over 6,000 inhabitants in about 417 households were connected to these microgrids. Ghana's central government in collaboration with the World Bank funded the development of these projects through the Ghana Energy Development Access Project (GEDAP) which were officially turned over to the Volta River Authority, the main government agency responsible for generating and supplying electricity.

Private investment in microgrids is limited as the systems are developed and operated by a private entity without government funding. Only a few companies have ventured into this business. Black Star Energy, a Ghanaian company specializing in microgrid development, was the first private entity to be licensed to operate solar powered microgrids. The company now operates 15 projects in the Ashanti region with ongoing developments in nearby regions. The four communities being considered for microgrids are Aniatentem, Affulkrom, Amanhyia, and Odumasi, all in the Bekwai Municipal Assembly.

Capacity and Reliability

Government-managed solar PV microgrids were constructed with a combined energy capacity of 650 kWh/day. The microgrid configurations included a backup diesel generator to provide supplementary power. These DG microgrids are low voltage consisting of a three-phase backbone feeder with a single-phase lateral that connects single loads at the customer's premises. In addition, the systems are configured with a dedicated feed for high efficiency light emitting diode (LED) street lamps throughout the villages. Households are connected and paid in accordance with a contract tariff class which is determined by their monthly electricity consumption.

Private model microgrids use solar PV and battery storage. The capacity of the systems ranges from 5 kW to 20 kW with a sizeable development pipeline. They are designed with the ability to be scaled to meet household and commercial electrical demand.

The electrical capacities of microgrids developed by both the government and private operators were below consumer expectations. It was revealed that the electricity supplied to end-users could not support the load of some of their home appliances. This was more prevalent in communities with government-managed microgrids. Private model customers who could afford to pay for access to additional power for their home appliances were supplied with more electricity. Our results indicated that private model consumers were more satisfied with the reliability of the electricity supplied than those under the government model. The reliability of the solar PV and battery systems were somewhat affected by weather and technical challenges in the island communities.

Technologies

To comply with the licensing provisions, applications for microgrids must include information about the technologies to be used, installation capacities, and the experience the company has in microgrid deployment. The technologies deployed by the government and privately owned businesses are similar as they use solar PV designs based on semiconductor materials that directly convert sunlight to electricity. The electricity supplied is direct current (DC) which reduces costs and losses as AC/DC inverters are not required.

Systems under the government model sell electricity on a pre-paid basis using an energy daily allowance (EDA) with smart-metering to eliminate defaults. Consumers cannot acquire electricity unless it is approved by a community-based operator.

However, the system is unable to automatically detect illegal connections and notify authorities which decreases tariff collections. It relies on the manual checks by a community-based operator.

Unlike government microgrids, private systems provide their consumers with opportunities to request and remotely purchase electricity via a mobile transaction whenever more power is required. Users can also notify operators via text message of faults, though operators normally perform weekly system checks. The detection technology used by private microgrids makes it impossible for users to engage in illegal connections.

Tariffs and Rates

The government is bound by its uniform national tariff (UNT) which regulates fees charged to residents of rural island communities that use electricity from solar PV microgrids. The UNT mandates that all domestic customers pay the same tariff for electricity. The UNT rates are important to end-users in remote communities served by microgrids. The island communities using the government model are covered by the UNT and their subsidized tariff rates are relatively low. The most a typical household spends for electricity is between 25 and 35 Gh¢ monthly (1 Ghanaian Cedi = 0.18 USD) with some households spending as little as 10 Gh¢ per month. The median monthly income in Ghana is 4,906¢ GHS ($853 USD).

Consumers served by private developers do not enjoy government-subsidized tariffs. Microgrids developed and operated by private entities such as Black Star Energy have tariffs higher than those charged by the government. Tariffs are guided by the levelized cost of energy and designed to recover all costs invested in developing the microgrid. Under this system household spending for electricity ranges from 35 Gh¢ to 50 Gh¢ per month.

IMPACT OF MICROGRID DEVELOPMENT

Analysis of data from households under both government and privately managed models reveals that there have been positive improvements in people's lifestyles in the communities of Pediatorkorpye, Atigagorme, and Kudorkope. These impacts were economic, social, and educational.

Impact on Expenditures for Fuel

The total and average fuel consumption and expenditures for fuel declined due to the community microgrid projects. The data and observations made from discussions with community members indicate that a considerable amount of funding has been expended to secure indigenous energy sources (e.g., kerosene, dry cell batteries) for lighting. To estimate the household expenditures, the avoided costs of fossil fuel purchases and travel expenses to the market town to purchase those fuels must be included. The average expenditure on fuel was much greater in the government communities. Village residents could only travel by motor boat to purchase fuel, at a cost of at least 10 Gh¢ per person for a round-trip journey on market days. The average monthly expenditure on fuel per household was about 30 Gh¢ to 40 Gh¢. The situation with communities under the private model is similar but they could travel any day of the week to secure fuel for lighting.

Analysis of the data has established that households under both the private (57%) and government (62%) models reduced their expenditures for fuel. Those under the private model saved about 5 Gh¢ to 7 Gh¢ per month while those under the government model saved between 5 Gh¢ and 10 Gh¢ per month.

Impact on Women

When the welfare of women in rural communities improves so does the welfare of their entire family. The surveys indicated that microgrid projects have improved the lives of local women and they are happier. It was found that women feel more secure with solar lighting at night which provides better illumination and facilitates mobility. Women generally indicated that they had more free time as they were relieved from the previous duties of finding fuel for lighting and other activities. They now use this time to engage in commercial activities and leisure. While few women participate in commercial endeavors, those who do sell fish, pure water, and ice cream. They have indicated that their incomes have increased by 40 Gh¢ to 60 Gh¢ per month as a result of having access to the electricity supplied by the microgrids.

Impact on Education

The impacts of microgrid development on education in communities under the government model were more pronounced than in communities under the private model. All four communities have schools but student attendance and performance were lower prior to the development of the microgrids. School attendance has since improved. Daily learning hours increased from an average from two to three hours to five to six hours per day—almost twice the previous number of hours.

The impact of solar PV microgrids on education under the private model was less than in the communities under the government model. The main reason is that there were no schools situated in those communities and children must travel long distances to attend school. Student attendance had not increased as compared to the communities under the government model. However, it was reported that learning hours had improved to an average of four to six hours per day.

CONCLUSIONS AND RECOMMENDATIONS

For Ghana's rural communities, solar PV offers a renewable energy solution for generating electricity that is a vast improvement over the use of fossil fuels. To assess the viability of solar PV microgrids in Ghana, we initially provided an analysis of the data obtained from the surveyed companies and communities. The literature review indicated that the regulatory environment is crucial for microgrid viability. Ghana has policies to promote, develop, and use microgrid systems in off-grid communities and regulate private sector engagement in microgrid development. Licenses are required to install and operate microgrids. Regulations apply to microgrids with DG capacities from 100 kW to 1 MW. Ghana has both private and government models for ownership and operation of solar PV microgrids.

Conclusions

Our analysis revealed interesting findings regarding the capacities and reliability of private and government solar PV microgrids serving the communities studied. Irrespective of ownership, generation capacities are unable to meet consumer demand because capacities were unable to handle the loads of some residential appliances. Unlike the government model, private companies allowed customers to purchase additional electricity according to their needs if they had the ability to pay. This is contrary to the view that additional electricity generation is not required in rural communities; more capacity is actually needed. Solar PV microgrid capacity and supply must be adequate to continuously meet consumer loads.

It is difficult for system operators to mandate which appliances consumers use and when they can use them. To be successful, the government model must mimic the private model by improving flexibility and allowing consumers to purchase power as needed. Interestingly, despite having diesel generators to provide backup power, the government model was less reliable than private systems that used only solar PV with battery storage. The backup diesel generators were unable to resolve reliability problems amidst weather and technical challenges. This suggests that solar PV microgrids may not need diesel backup generation to maintain reliability. Good management and assessment of environmental and technical challenges can make solar PV microgrids more reliable.

Generally, the government-owned business model uses older technology. While project development approaches are similar, the privately owned business model was found to provide better performance, better reliability, improved monitoring of illegal activities, better fault reporting, and greater revenues from sales. The private model avoided power losses, detected illegal activities, and allowed customers to purchase power as needed. The private model for solar PV microgrids is more customer-focused, and responds to the issues of profitability, sustainability, and viability.

Customers under the government model enjoyed subsidies under UNT yet costs were not offset. Tariffs under the private model were higher than those under the government model. However, despite the subsidies, government customers had more power supply and reliability problems.

The socioeconomic benefits of microgrids that provide solar-generated electricity to rural communities are important. Indoor air quality is improved as there is less reliance on kerosene as a fuel source. Other benefits include cost savings on fuel purchases, improvements in the welfare of women, and educational benefits such as increased hours of learning. With solar-generated electricity, household expenditures were much lower when compared to previous expenditures on kerosene and dry cell batteries. Cost savings were among the benefits for households supplied with electricity by both privately operated and government-operated microgrid systems. Women now enjoy improved security and have more free time, which is used for commercial and leisure activities. Remarkably, commercial activities performed as a result of greater access to electricity increased incomes by about 40 Gh¢ to 60 Gh¢

monthly. Concerning education, students in the various communities under the two models had increased hours of learning due to solar PV microgrids. Attendance improved at schools under the government model.

Community, environmental, and geographic factors have beneficial effects on the acceptance and adoption of solar PV microgrids. For instance, the analysis showed that in some rural communities prior to the operation of the microgrids, electricity was expensive because of travel costs associated with transport by boat to buy fuel. The transportation costs increased the expense of fuel acquisition. The development of solar PV microgrids brought relief, saved time, avoided risks associated with river transport, and eliminated transportation fees. In one community, morning weather patterns affected the supply of electricity from the solar PV microgrids, emphasizing how weather can impact the viability of solar PV microgrids. Consequently, to fully assess the benefits and viability of solar PV microgrids, the community, environmental, and geographic aspects must be considered.

Solar PV microgrids in rural communities are economically and technologically viable using the private business development model. The viability of microgrids depends on adopting the appropriate technologies for development and operation considering metering systems, flexibility in demand, power loss prevention, revenue generation, detection of illegal activities, and reporting of faults. Other factors may also impact the viability of a microgrid.

Recommendations

Private participation in solar PV microgrids is generally preferable and must be promoted rather than the governmental model. The private model takes contingent steps with the help of technology to avoid risks and ensure reliability, profitability, and sustainability. The government should refrain from the policy of extending subsidies. This is because the government may fail to make the necessary transfer to the private entities which impacts the operation of private companies and electrical power reliability. Furthermore, in every solar PV microgrid project, emphasis should be placed on providing the capacity consumers require. To be successful microgrid operators must supply power to match customer loads. This necessitates predictive forecasting of energy demand. We recommend that a cash flow analysis be performed for each project to establish the profitably and financial sustainability of the various microgrid alternatives.

Providers should be mindful that provision of electricity has beneficial social and economic impacts on communities and creates commercial initiatives. Developers and researchers should be concerned with the present need for electricity in rural areas and its multiplier effects such as lifestyle improvements, new business development, and expanded economic activities.

ACKNOWLEDGMENTS

This chapter is an updated version of an article entitled "Ghana's Transition to Renewable Energy Microgrids: An Assessment of Ownership, Management and Performance Dynamics" that was originally published in the *International Journal of Strategic Energy and Environment Planning,* Volume 2, Issue 3, Stephen Roosa, Ph.D. (editor) in March 2020. The editor extends thanks and credit to the Association of Energy Engineers, Atlanta, Georgia for use of previously published material and to the authors who allowed the article to be republished.

REFERENCES

1. United Nations (2018). Sustainable development goals report 2018. New York.
2. World Bank Group (2018). Energy home – overview. http://www.worldbank.org/en/topic/energy/overview#3, accessed on 24 November 2019.
3. International Renewable Energy Agency (2016, September). Solar PV in Africa: costs and markets. https://www.irena.org/publications/2016/Sep/Solar-PV-in-Africa-Costs-and-Markets, accessed 9 August 2019.
4. Energy Sector Management Assistance Program (2016). Ghana: microgrid for last-mile electrification. Exploring regulatory and business models for electrifying Lake Volta Region. *The World Bank.* Knowledge Series.
5. Arranz-Piera, P., Kemmansuor, F., Darkwah, L., Edjekumhene, L., Cortes, J., and Velo, E. (2018). Microgrid electricity service based on local agriculture residue: feasibility study in rural Ghana. *Energy*, 153, pages 443–454.
6. Gollwitzer, L., Ockwell, D., Muok, B., Ely, A., and Ahlborg, H. (2018). Rethinking the sustainability and the institutional governance of electricity access and microgrids: electricity as a common pool resource. *Energy Research and Social Sciences*, 39, pages 152–161.
7. Comello, S., Reichelstein, S., Sahoo, A., and Schmidt, T. (2016). Enabling microgrid development in rural India. *World Development*, 93, pages 94–107.
8. Bhattacharyya, C. and Palit, D. (2016). Microgrid based off-grid electrification to enhance electricity access in developing countries: what policies may be required? *Energy Policy*, 94, pages 166–178.
9. Bhattacharyya, C. (2014). Viability of off-grid electricity supply using rice husks: a case study from South Asia. *Biomass and Bioenergy*, 68, pages 44–54.
10. Szabo, S., Vallve, V., Ackom, E., Lazoponlou, M., Solano-Peralta, M., and Moner-Girona, M. (2018). Electrification of sub-Saharan Africa through PV/hybrid microgrids: reducing the gap between current business models and on-site experiences. *Renewable and Sustainable Energy Reviews*, 91, pages 1,148–1,161.
11. Painuly, J. (2001). Barriers to renewable energy penetration: a framework for analysis. *Renewable Energy*, 24, pages 73–89.
12. Knuckles, J. (2016). Business models for microgrid electricity in base of the pyramid markets. *Energy for Sustainable Development*, pages 67–82.

11 Local Energy Supply Possibilities—Islanding Microgrid Case Study

István Vokony, József Kiss, Csaba Farkas, László Prikler, and Attila Talamon

INTRODUCTION

In low-voltage distribution systems, some networks are not operated and maintained by an authorized distribution system operator (DSO). These subnetworks are usually connected to the synchronous system at a single point. The accountable measurement is resolved exclusively at this specific connection point. After that point, submeters serve as a basis for accounting among users. At these network endpoints voltage-related problems arise frequently since the grid is often in poor condition after the connection point. There are two common solutions to this problem: 1) eliminate the subnetwork and substitute it with a conventional system from the medium-voltage (MV) to low-voltage (LV) transformer endpoint; or 2) provide an operational microgrid island system for consumers at the statutory level. In this chapter, electrical service quality is analyzed, construction and operational requirements are cited, and through on-site measurements, a renewable energy mix optimizer application is introduced. Based on the measurements and operational experience, a calculation methodology is established to compare a traditional solution with the island operation possibilities. By comparing network development costs and islanding solution costs, an optimal weight factor can be defined which can be used as an effective investment decision support parameter.

Over the last century, the world's electricity supply has been provided primarily by centralized power systems using conventional power plant technologies—combusting fossil fuels (coal, natural gas, etc.)—or nuclear energy or renewables such as large hydroelectric power plants [1,2]. In these systems the power plants, producing hundreds of megawatts or even gigawatts of power, transfer electricity to distribution networks and then to the end-users through a transmission network. However, in the last few decades, production has shifted somewhat to the consumer side with much smaller energy generation units [3]. An ever-increasing number of these use renewable energy. Another major change is that information technology (IT) systems are now linked to electricity production and supply support processes. Rather than rebuilding the aging network, new island microgrid systems using distributed power generation and energy storage technologies should be developed [4].

There are special locations (e.g., on remote islands, in mountainous areas) where the only way to provide electrical power is to establish local energy systems. The local island microgrid solution can be competitive in cases where a synchronous network is available, but new approaches are needed for economic reasons. One catalyst for new approaches to providing electricity is the growing renewable generation industry.

INTERNATIONAL OVERVIEW

According to the International Energy Agency, over one billion people live without electricity, accounting for 17% of the Earth's population. Almost all of these people live in developing countries in Asia or Africa, where local renewable electrical energy generation can often be the most economical solution.

Many available electrical system standards lack explicit specifications for the power quality of off-grid systems (i.e., island microgrids). We can use the protocol for networked systems, though there are partial regulations in European Norms (EN) 61000 standards [5,6]. Power quality, however, holds special importance in partial systems, because the intermittent behavior of the weather-dependent renewables (sun and wind) results in significant power fluctuations which also impacts supply voltages. Inverters operating in island mode must also comply with the Institute of Electrical and Electronics Engineers (IEEE) 519-014 standards [7,8].

The local grid definitions developed by the Smart Grid Interoperability Panel (SGIP) [9] provide several definitions for networks that can be operated in island mode:

- According to the CIGRÉ (International Council on Large Electric Systems), *microgrids* are networks with consumers and distributed generation units that can be operated independently of the main electrical grid (see Figure 11.1). These units connect to one single central control unit. The definition is congruent with that given by the U.S. Department of Energy. The SGIP notes that these definitions disregard off-grid systems that are grid-independent or grids not connected to a main network [9].
- Given the above, a new definition for *microgrid* is proposed: a distribution grid that consists of consumers and distributed power generators and can be operated independent of the electricity system by islanding at one single central regulator (see Wissner et al. [10]).

The SGIP also specifies smaller units as *nanogrids* or *picogrids* [9]. Picogrids are device level systems. The nanogrid lacks a precise definition; it usually indicates a local energy community smaller than a microgrid which can be operated in island mode independent of the main grid. The SGIP has noted the inaccuracy of the definitions of a nanogrid [9]. Almost all the definitions concur that the nanogrid is a direct current (DC) system. In most cases, a nanogrid is defined as a single building or a building complex, which consumes only a few kW.

Definitions for *islanded operation* of a local grid include that of the International Electrotechnical Commission (IEC) which states it is a specific part of the electricity

Local Energy Supply Possibilities

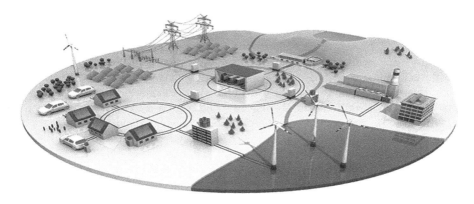

FIGURE 11.1 Potential smart microgrid [6].

system that is not connected to the interconnected system but is nevertheless energized [9]. *Island* in a power system is "a portion of a power system that is disconnected from the remainder of the system but remains energized" [11].

There are several forms of islanding as detailed by IEEE [8]. A common feature is that all assume a distribution network that is not electrically connected to the main grid due to intended or unintended (i.e., due to a fault) disconnection. An islanded microgrid grid can be characterized based on its size according to IEEE [8]:

Local island: producers and consumers located within a facility/building.
Secondary island: several producers and consumers connected to the secondary winding of a MV/LV transformer.
Side island: consumers and producers connecting to one side of a distribution network.
Section island: one branch with its consumers and production units.
Substation bus island: a part of a network supplied by a substation bus.
Substation island: a whole district supplied by the substation.

In Kempener et al. the off-grid networks are defined as those that are not used or do not depend on the electricity provided by the interconnected power grid [12]. Within these networks, we distinguish those using renewable energy and traditional ones which use primarily diesel generators.

The voltage quality supplied by the islanding system—unless indicated differently—must comply with current regulations for a low-voltage connection point. The voltage at the point of connection shall be within a ±10% range over the course of a one-week measurement period under normal operating conditions in 95% of the averaged values for any ten-minute period [13]. All ten-minute averages of the one-week measurement period must fall within ±10% of the nominal value. The maximum voltage increase may exceed 115% of the nominal voltage; at the same time, the voltage drop may not reach 80% of the nominal average voltage for any one-minute interval.

In the case of islanding supply, it is highly probable that the condition "there is electricity in its surroundings" is unmet, which may exempt the service provider from troubleshooting obligations. The lack of service is more likely caused by weather conditions (local production failure) than by a fault in the licensee's machinery.

Any possible operation after interconnection and disconnection (such as islanding) must occur according to the IEEE 1547 standards. Bhatacarya et al. mention that the relevant standards do not include specifications for off-grid networks, but suggest that they be more permissive than the requirements for the electricity grid [14]. This contradicts Bollen et al. who state (although it is not explicitly about the off-grid network, but only about microgrid operating in islanding) that a system operating in island mode may not have worse indicators than what is established as quality requirements for interconnected electrical grids by the EN 50160 standards [15]. In fact, it is assumed that quality indicators may show even better results than those in the distribution network.

Guaranteed Service Levels

While there are guaranteed service level requirements for off-grid systems, the issue is relevant. The Council of European Energy Regulators (CEER) urges meeting the Fit Compliance team required regulations [16,17]. For example, neither the requirements of the Hungarian Energy and Public Utility Regulatory Authority nor of the UK's Office of Gas and Electricity Markets (OFGEM) for the guaranteed service level concerning an islanding system are, as yet, determined [18,19]. However, they are similar to the Microgeneration Certification Scheme that states that off-grid systems can receive feed-in tariffs if they meet the specifications for microgrids [20,21]. In such cases, off-grid systems are to be treated as any other casual system.

It is likely that the existing guaranteed service level requirements will be applied in a similar manner to that of an interconnected system, but doing so may result in additional costs [22–24].

Construction Aspects

During the construction stage of a microgrid project, many components of an islanding system need to be considered (e.g., on-site measurements, location and size of capacitors, voltage regulators, reactors, protection devices, and transformers) [25, 26]. Planning for the anticipated consumer loads and their load characteristics needs to occur prior to construction and reevaluated annually in the light of any proposed changes. By determining the characteristics of the production units and defining black start capability, the behavior of the islanding microgrid can be defined when frequency and/or voltage changes occur.

The limit values of voltage asymmetry must be identified, stability limitations must be specified, and the specifications for the protection devices must be provided. The potential of future microgrid expansion should also considered.

Transient voltages can occur when large loads (e.g., motors) are started; an action plan is needed for this situation (e.g., apply soft start capabilities for motors). This is particularly important if the island microgrid is weaker than the distribution network, as voltage dips due to the motor starting will be more significant. System grounding (earthing) is also important; all pertinent guidelines for interconnected networks should be followed.

If there is inadequate generation capacity in the island, appropriate load shedding strategies must be implemented [27]. If there is an emergency diesel generator, switchover specifications are needed.

Because an island microgrid is a smaller system than a distribution network, certain problems that may arise in the interconnected power system may be more common when islanding:

> *Load asymmetry*: careful attention must be paid to the proper phase assignment; in the case of single-phase consumers, a blown fuse might cause significant asymmetry.
> *Inrush current*: there may be smaller nominal power transformers in the microgrid when operating as an island; in these instances, the inrush current can be mitigated by properly sizing the system. Furthermore, IEEE mentions an issue with motors in the island. As laboratory measurements prove, when switching from one source direction to another, the motors have an inrush current when started [8].
> *Power outages*: for lighting devices, it is important to have a continuous supply of electricity in the island. As in the case of power outages, depending on the technology, it may require as much as a minute to recover the original operation of the light source [28]. Depending on the type of the island, it may be necessary to provide emergency lighting with a dedicated electrical supply.
> *Frequency control*: the primary controller must operate in isochronous mode. If there is more than one generation unit, regulation requirements must be determined. According to IEEE, the so-called V/Hz ratio should also be monitored. This value is provided for each device (transformer, motor, etc.) and operating at higher values may cause energy losses or damage due to heating [8].

Employee training is important to keep the operating company's technicians, mechanics, and staff up to date on system maintenance and operations (how to manage black starts, what protection schemes should be used, etc.).

Planning Principles

For an off-grid network using renewables, sizing must be based on the peak power consumption and on the annual electrical demand [29]. This can be challenging since supply power is to be provided for areas previously not electrified, yet no measurement results are available to gauge the requirements. In such cases, surveys can be used to canvas consumer habits and needs of similar but previously electrified districts [30].

An example from Bosnia and Herzegovina (Figure 11.2) shows a typical consumer curve for households in sparsely populated rural areas [31]. In addition to the daily consumer profile, it is also necessary to consider seasonal variations (see Figure 11.3).

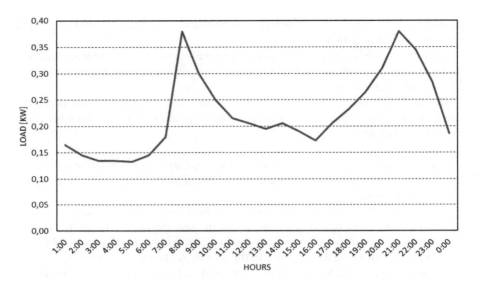

FIGURE 11.2 Average customer profile in one-hour resolution.

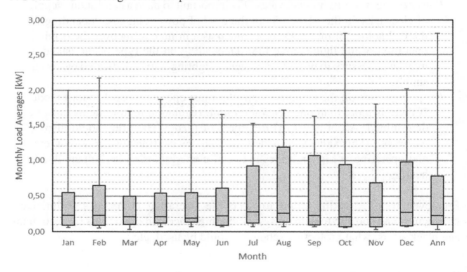

FIGURE 11.3 Seasonal change in average consumption.

To avoid or manage these volatilities, one solution is to use energy storage technologies, typically battery energy storage.

Operational Aspects

IEEE defines the operating requirements for distributed resource island systems [8]. These include: 1) meeting the real and reactive power demand of the consumers; 2) keeping the voltage and frequency close to their nominal values (this does not apply to long-term islanding operation but to those parts of a microgrid system that become islanded); and 3) providing the system with proper control capabilities. To have controlling capability, a generation unit that can be operated outside the ranges defined in IEEE 1547 is needed (i.e., if a motor requires a large amount of reactive power to start, a generating unit must be available with sufficient reactive power capacity) [32]. Fluctuations in load dynamics have to be adequately followed. This can be achieved by incrementally ramping loads, load management, or load shedding.

It is necessary to provide adequate dynamic characteristics to follow the potential changes. Such transient stability requirements arise when there is a sudden change in load, a production unit outage occurs, or faults occur inside the island.

Protection devices different from the conventional ones are required (e.g., the short circuit current is usually lower, so the conventional protection devices must be tuned to these new requirements).

According to IEEE requirements, microgrids operating in island mode must also have adequate dynamic characteristics [8,33]. They must be sensitive to sudden changes in voltage and frequency and capable of reacting properly to these changes [34].

If relevant sources are missing, the dynamic behavior of the islanding operation's voltage and frequency control device should be monitored by on-site measurements [35].

Another important issue is the synergy of the island's frequency and voltage stabilizer device (for small systems this means a battery inverter) with other similar devices of the island (photovoltaic inverters) [36]. Current international research efforts are focused on coordinating the maximum power point tracking (MPPT) functions, battery charging and lifecycle management functions, and islanding regulations [37].

LOCAL SOLUTION: CONTAINER MICROGRID

Next, a specific application is considered which is called a container microgrid. This provides a detailed and measurable example of microgrid considerations. The primary objective of container-scale microgrids is to have access to electricity in areas where no synchronous network connection is possible because of high investment costs and/or a low rate of return. A *mobile power plant* is one solution to this problem. It may be used in army field practices, telecommunication equipment operations, humanitarian missions, crisis or emergency situations, and in assisting reconstruction or black starts.

For this application, a specific container microgrid was made available to be examined and measured (see Figure 11.4). Based on a detailed analysis a decision support application was created. The container was designed to operate as a microgrid in island mode. It generates electricity mainly from renewable energy sources (solar and wind power). This energy can be stored in batteries which can be used for buffering. When the battery charge level falls below a specified value (in this case 20%), the automated system switches on a diesel emergency generator.

The main feature of the power plant is its mobility. All devices are placed in a 20-foot long marine container. The container forms the structure's base while in operation.

FIGURE 11.4 Container microgrid when installed [38]. (Source: E.ON, Hungary grid innovation department, 2016.)

Because of the standard size, it can be easily replicated and transported by boat or truck. With a unique container being available for study, on-site measurements the operational limits of several scenarios could be tested to assess the reliability of island operation.

Container Microgrid Assessment

Small-scale microgrid island systems often utilize renewable energy sources such as solar photovoltaic (PV) modules. These devices produce DC electricity; inverters are needed to create three-phase alternating current (AC) network. To deal with the load fluctuations and asymmetries as previously described, an adequate inverter design must be used.

We used a number of investigative approaches to assess the transient behavior of a small microgrid operating in island mode. Based on our results, on-site measurements were performed for the energy container. In our investigations, various asymmetrical and symmetrical load configurations were used to simulate consumer behaviors. During the investigation period, the inverters maintained the voltages and the frequency within the permissible ranges (50 Hz ± 0.05% for frequency), regardless of the asymmetries or load fluctuations.

The electric parameters of the three-phase inverter that supplied this small-scale islanding system are described as follows:

- nominal load capacity: continuous load 3.5 kVA
- overload capacity: up to 30 minutes with 4 kVA
- short-term overload (power boost) up to five seconds with 10.5 kVA

Local Energy Supply Possibilities

- output voltage: sinusoidal, 230V ±2%
- frequency: 50 Hz +/- 0.05% (49.975 Hz to 50.025 Hz)
- total harmonic distortion calculated for voltages (THD_u): < 2%

To investigate the transient behavior of the islanding system and the feeding inverter, artificial loads were used to simulate the load behavior of a residential consumer. The switching events used to examine the transient behavior of the system are listed below and the load changes in each phase are shown for the sample period 11:42 to 12:00 (see Figure 11.5). Only devices that could be found in a typical household were used for the simulation.

- In the phase (denoted as "S" in Figure 11.5), a heating fan with temperature control was operated that automatically cycled off when the required temperature was satisfied (± 2kW changes on the power curve).
- The only change in the apparent power measured in phase S was the activation of a vacuum cleaner.
- The power peaks that took place simultaneously in all phases occurred when the 3 × 980W consumer was energized.
- Specific load peaks in phase T were caused by turning the microwave oven on and off.
- At 11:56, phase S was switched off by the inverter for overload protection. During the measurement periods, the root mean square (RMS) values, time functions of the phase voltages, phase currents, and frequency of the consumer island fed by the container were recorded.

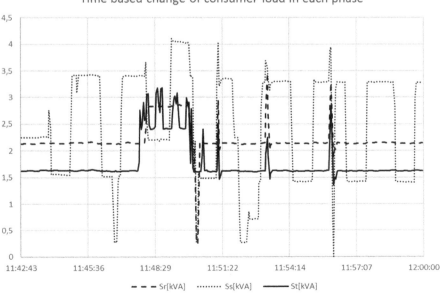

FIGURE 11.5 Changes in consumer loads in each phase.

Figure 11.5 depicts the resulting load fluctuations in each phase for the sample period. Figure 11.6 illustrates the phase voltages maintained by the single inverter feeding the islanding system. Generally, the inverter managed to cope with sudden load changes and maintained voltages within the permitted ranges. The single exception happened at 11:56 when phase S was switched off by the inverter for overload protection.

To conclude the on-site measurements, the annual energy consumption, maximal power, plus the possible load asymmetries and fluctuations must be accounted for when designing an islanding system. This is especially true if the consumer group consists of only a few households. Within these parameters an appropriate solution can be established that is capable of supplying the microgrid's consumers with proper voltage quality.

DECISION SUPPORT FOR SOFTWARE DEVELOPMENT

As previously explained, in certain cases establishing an islanded microgrid can be a valid alternative to extending a distribution grid. Next, a simple and easy-to-use decision support tool developed by the authors of this chapter is introduced for use by interested professionals.

For this example, an elaborated Microsoft Excel-based decision support software provides a site-specific technical proposal for implementation based on network parameters, consumer data, conventional network development, or alternative solutions (e.g., energy containers). The application consists of several imbedded worksheets (see Figure 11.7). Worksheets labeled "IN" contain input data that the user needs to enter or download from another (e.g., measurement) database in advance. These worksheets provide input data for worksheets that perform the substantive calculations. The worksheet *tag* indicates which input data sheet contains input

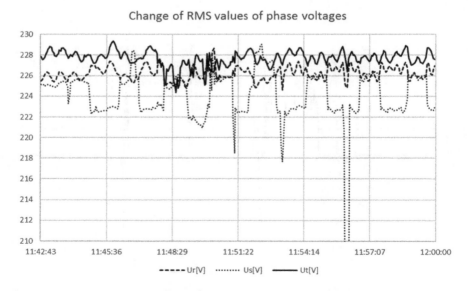

FIGURE 11.6 Voltage fluctuations due to time-varying consumer loads.

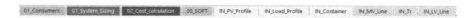

FIGURE 11.7 Overview of worksheets for decision support Excel workbook.

information for the spreadsheet. These worksheets mainly include profiles, type information, and specific costs. Cells with input data are highlighted on each worksheet (Figure 11.8).

Worksheets whose name begins with Arabic numerals contain diagrams that help to evaluate the results. The *planning steps* follow each other according to the numbering of the worksheets. The selection of the components actually used and the technical specifications are indicated on these worksheets.

Results, including output data, are shown with a gray background (Figure 11.9). In some cases, the value of some input data can also be changed by *sliders* or *scrollbars* (Figure 11.8), which allow a quick overview of the effects of the various changes.

COST OF NETWORK DEVELOPMENT ALTERNATIVES

In addition to the technical solution of a container supply, it is also possible to estimate the cost of the conventional network alternative (wire, earthworks, transformer, etc.) This estimate accounts for three possible cost components: to build a medium-voltage (MV) network, to install a MV/LV transformer, or to build a low-voltage (LV) network. Cost estimates are based on the knowledge of specific costs. After selecting the appropriate wire types and setting the required lengths, the cost of constructing the overhead transmission lines or cables can be estimated. For wiring, it is possible to specify a mandatory amount for the transformer. The costs for this (in accordance with the distribution rules) normally belong to the DSO. For this analysis the possibility of cost sharing between the customer and the service provider is not considered.

Comsumer data				Default values
Annual Consumption	[kWh]	3600		
Max. power demand per phase	[kVA]	2.3		
Number of customers	[pcs]	2		
Asymmetry factor	[%]	51	< >	100%
Load simultaneity factor	[%]	100	< >	100%
Max. permissible daily consumption	[kWh/day]	16		
Simultaneity factor of the daily consumptions		100	< >	100%
Geographical location (area, region)		Budapest	Choose from the list!	100%
Proportion of the consumption supplied by PV	[%]	100	< >	100%
Max. time of operation without production	[day]	7		

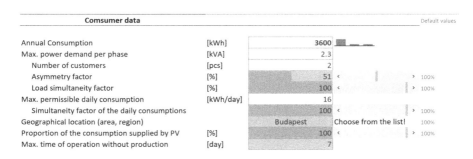

FIGURE 11.8 Example of cells with input data.

Results - System components			
PV peak power	[kWp]	8.0	PV power rating is higher than that of the inverter!
Inverter power, 3 phase	[kVA]	6.9	
Required storage capacity	[kWh]	251.5	

FIGURE 11.9 Example of cells displaying computational results.

To examine the islanded system for each customer, the decision support Excel workbook was used. In the created worksheets, the predefined equations calculate the technical parameters of the supply of the insulated customer or consumer groups by different inputs. As its name suggests, the file is not a precision scaling program. It is intended to approximate the parameters and costs of the microgrid and support the process of deciding whether an islanded system is viable.

Essential data for load model:

- annual consumption (kWh)
- maximal demand (kVA)
- permitted daily maximum consumption (kWh)
- annual consumption profile, monthly breakdown

Adjustable parameters of the energy container:

- inverter rated power (kVA)
- PV rated power (kWp)
- battery capacity (kWh)
- costs (Ft)

After specifying the parameters, the application will perform the scaling of the system and specify the data for the following items to be applied:

- *PV rated power (kWp)*: determined by monthly production and consumption (in the case of multiple consumers, the result of each profile).
- *Inverter rated power (kVA)*: the maximum single-phase power requirement is three times the one phase. Assuming a symmetrical three-phase layout, the application can handle the asymmetric load distribution as well.
- *Storage capacity (kWh)*: the capacity of containers is determined by approximating the table based on the maximum production time (eight days) and maximum allowable daily consumption.

Based on the determined values, it is possible to calculate how many sections of a particular container type are required to be installed. This can be illustrated by developing two diagrams of the intended system: 1) one to describe the monthly electrical demand of consumers and the expected production of the solar PV system; and 2) another showing monthly production and consumption, maximum output from overproduction, and corrected consumption using the previous maximum storage to estimate the unmet system demand. The second diagram can show an estimate of the long-term gap that can be overcome by providing electrical storage.

The technical calculations are an invaluable decision support tool. In addition, an economic comparison can be made using the application to estimate the network costs (transformer, MV and LV network), the wire types, and their predefined costs. The network licensee and customer pricing can be specified. The tool can be also used to estimate the economic return of the comparative solutions over a 20-year period from the perspective of the network licensee.

SUMMARY

Islanded operation of a local microgrid often refers to a specific part of the electricity system that is not connected to the interconnected system. There are a number of forms of island operation. Any possible island operation after interconnection and disconnection must occur according to the IEEE 1547 standards. Since an island microgrid is much smaller than a distribution network, problems that occur in interconnected power systems may be more common when islanding.

Providing a quality electrical supply in a microgrid island operation requires a complex analysis that has legal, technical, and economic considerations. Based on tests with a sample container microgrid, this chapter defines the key parameters needed for a technical and economic analysis, and suggests decision-makers use a decision-support spreadsheet for their assessments. Using this tool, the example island-based solution becomes economically feasible when compared to the investment costs of establishing a power supply network with public transmission lines and a conventional transformer.

The voltage quality requirements for islanding systems are the same as for low-voltage connection points. The current requirements for service continuity would be difficult to apply to islanding systems. To resolve this, it is necessary to update the relevant regulations, including the attributes of the guaranteed service level. For voltage quality, the issue of harmonic distortion is of utmost importance for microgrid electricity production units and energy storage systems that can be connected to the community internal network using inverters. These inverters behave as non-linear generators, and because only a fraction of the usual short-circuit power is available in the island, the voltage distortion they cause will be significantly greater than that of the interconnected network.

Another important issue is the cooperation of the island's frequency and voltage stabilizer devices with the other similar devices of the island (e.g., additional solar PV inverters). In the case of frequency control, the primary controller must operate in isochronous mode; if there is more than one production unit, it must also be determined how they are regulated.

ACKNOWLEDGMENTS

This chapter is an updated and adapted version of an article entitled "Local Energy Supply Possibilities—Islanding Microgrid Case Study" that was originally published in the *Energy Engineering,* Volume 115, Issue 6, Steve Parker (editor) in 2018. Reprinted by special permission of The Fairmont Press, Inc.

REFERENCES

1. Razmara, M., Bharati, G., Hanover, D., Shahbakhti, M., and Robinett, R. (2017, October 1). Building-to-grid predictive power flow control for demand response and demand flexibility programs. *Applied Energy*, 203, pages 128–141. https://www.sciencedirect.com/science/article/pii/S0306261917307936, accessed 4 April 2018.
2. Yan, J., Zhai, Y., Wijayatunga, P., Mohamed, A., and Campana, P. (2017, September 1). Renewable energy integration with mini/micro-grids. *Applied Energy*, 201, pages 241–244. https://www.sciencedirect.com/science/article/pii/S0306261917307122, accessed 9 January 2020.

3. Balcombe, P., Rigby, D., and Azapagic, A. (2015, October 1). Energy self-sufficiency, grid demand variability and consumer costs: integrating solar PV, Stirling engine CHP and battery storage. *Applied Energy*, 155, pages 393–408. https://www.sciencedirect.com/science/article/pii/S0306261915007758, accessed 9 January 2020.
4. Xue, X., Wang, S., Sun, Y., and Xiao, F. (2014, March 1). An interactive building power demand management strategy for facilitating smart grid optimization. *Applied Energy*, 116, pages 297–310. https://www.sciencedirect.com/science/article/pii/S0306261913009719, accessed 9 January 2020.
5. Janiga, P., Liška, M., Beláň, A., Volčko, V., and Ivanič, M. (2015). Power quality measurement in low power solar off-grid system with load control simulation. *Advances in Environmental and Agricultural Science*, pages 224–228. http://www.wseas.us/e-library/conferences/2015/Dubai/ABIC/ABIC-31.pdf, accessed 4 April, 2018.
6. 3M. Smart grid: connected, efficient and sustainable energy. http://solutions.3m.com/wps/portal/3M/en_EU/SmartGrid/EU-Smart-Grid, 4 accessed May 2018.
7. Algaddafi, A., Brown, N., Rupert, G., and al-Shahrani, J. (2016, February 29). Modelling a stand-alone inverter and comparing the power quality of the national grid with off-grid systems. *IEIE Transactions on Smart Processing and Computing*, 5(1), pages 35–41. http://www.academia.edu/34106575/_6_Modelling_a_Stand-Alone_Inverter_and_Comparing_the.pdf, accessed 4 January 2020.
8. Institute of Electrical and Electronics Engineers (IEEE). Guide for design, operation and integration of distributed resource island systems with electric power systems. http://ieeexplore.ieee.org/document/5960751/, accessed 4 April 2018.
9. Smart Grid Interoperability Panel (2016). Local grid definitions: a white paper, by the Smart Grid Interoperability Panel, Home, Building and Industrial Working Group. https://eta-intranet.lbl.gov/sites/default/files/SGIP_Local_Grid_Definition_2.pdf, accessed 4 April 2018.
10. Wissner, M. (2011, July). The Smart Grid – a saucer full of secrets? *Applied Energy*, 88(7), pages 2509–2518. https://www.sciencedirect.com/science/article/pii/S0306261911000602, accessed 4 April 2018.
11. Electropedia. International electrotechnical commission. http://www.electropedia.org/iev/iev.nsf/display?openform&ievref=603-04-46, accessed 4 January 2020.
12. Kempener, R., d'Ortigue, O., Saygin, D., Skeer, J., Vinci, S., and Gielen, D. (2015). IRENA off-grid renewable energy systems: status and methodological issues. http://www.irena.org/DocumentDownloads/Publications/IRENA_Off-grid_Renewable_Systems_WP_2015.pdf, accessed 4 April 2018.
13. Zubi, G., Dufo-López, R., Pasaoglu, G., and Pardo, N. (2016, August). Techno-economic assessment of an off-grid PV system for developing regions to provide electricity for basic domestic needs: a 2020–2040 scenario. *Applied Energy*, 176(15), pages 309–319. https://www.sciencedirect.com/science/article/pii/S0306261916306043, accessed 4 April 2018.
14. Bhattacharya, S. (2013, December). To regulate or not to regulate off-grid electricity access in developing countries. *Energy Policy*, 63, pages 494–503. https://www.sciencedirect.com/science/article/pii/S0301421513008215, accessed 4 April 2018.
15. Bollen, M., Zhong, J., and Lin, Y. (2009, June). Performance indices and objectives for microgrids. *20th International Conference on Electricity Distribution*. Prague, Check Republic. http://www.cired.net/publications/cired2009/pdfs/CIRED2009_0607_Paper.pdf4, accessed 4 April 2018.
16. Numbi, B. and Malinga, S. (2017, January 15). Optimal energy cost and economic analysis of a residential grid-interactive solar PV system- case of eThekwini municipality in South Africa. *Applied Energy*, 186, Part 1, pages 28–45. https://www.sciencedirect.com/science/article/pii/S0306261916314921, accessed 4 April 2018.

17. FIT Compliance Team (2016, May 10). Renewable electricity–feed-in tariff: guidance for licensed electricity suppliers, version 8.1. https://www.ofgem.gov.uk/system/files/docs/2016/05/fits_guidance_for_licensed_electricity_suppliers_v8.1_0.pdf, accessed 4 January 2020.
18. Liu, N., Tang, Q., Zhang, J., Fan, W., and Liu, J. (2014, September 15). A hybrid forecasting model with parameter optimization for short-term load forecasting of microgrids. *Applied Energy*, 129, pages 336–345. https://www.sciencedirect.com/science/article/pii/S0306261914005182, accessed 4 April 2018.
19. Ofgem (2014, December 16). Supplier guaranteed and overall standards of performance, statutory consultation and proposals. London, SW1P 3GE. https://www.ofgem.gov.uk/sites/default/files/docs/2014/12/gosp_statutory_consultation_.pdf, accessed 9 January 2020.
20. The Microgeneration Certification Scheme (2016). http://www.microgenerationcertification.org, accessed 4 April 2018.
21. Penava, I., Saric, M., Galijasevic, S., and Penava, M. (2014, May 10). Perspectives for off-grid renewable energy applications for rural electrification in Bosnia and Herzegovina. *14th International Conference on Environment and Electrical Engineering*. doi: 10.1109/EEEIC.2014.6835884. http://ieeexplore.ieee.org/document/6835884, accessed 4 April 2018.
22. Fares, R. and Webber, M. (2015, January 1). Combining a dynamic battery model with high-resolution smart grid data to assess microgrid islanding lifetime. *Applied Energy*, 137, pages 482–489. https://www.sciencedirect.com/science/article/pii/S0306261914004024, accessed 4 April 2018.
23. Li, Y., He, Y., Su, Y., and Shu, L. (2016, October 15). Forecasting the daily power output of a grid-connected photovoltaic system based on multivariate adaptive regression splines. *Applied Energy*, 180, pages 392–401. https://www.sciencedirect.com/science/article/pii/S0306261916309941, accessed 4 April 2018.
24. Amrollahi, M. and Bathaee, S. (2017, September 15). Techno-economic optimization of hybrid photovoltaic/wind generation together with energy storage system in a stand-alone micro-grid subjected to demand response. *Applied Energy*, 202, pages 66–77. https://www.sciencedirect.com/science/article/pii/S0306261917306207, accessed 4 April 2018.
25. Nan, S., Zhou, M., and Li, G. (2018, January 15). Optimal residential community demand response scheduling in smart grid. *Applied Energy*, 201, pages 1280–1289. https://www.sciencedirect.com/science/article/pii/S030626191730819X, accessed 4 April 2018.
26. Pearre, N. and Swan, L. (2015, January 1). Technoeconomic feasibility of grid storage: mapping electrical services and energy storage technologies. *Applied Energy*, 137, pages 501–510. https://www.sciencedirect.com/science/article/pii/S0306261914004036, accessed 4 April 2018.
27. Sigrist, L., Lobato, E., Rouco, L., Gazzino, M., and Cantu, M. (2017, March 1). Economic assessment of smart grid initiatives for island power systems. *Applied Energy*, 189, pages 403–415. https://www.sciencedirect.com/science/article/pii/S0306261916318372, accessed 4 April 2018.
28. Misak, S., Stuchly, J., Vramba, J., Prokop, L., and Uher, M. (2014). Power quality analysis in off-grid power platform. *Power Engineering and Electrical Engineering*, 12(3), pages 177–184. http://advances.utc.sk/index.php/AEEE/article/viewFile/988/973, accessed 4 April, 2018.
29. Bhandari, B., Lee, K., Lee, C., Song, C., and Ahn, S. (2014, November 15). A novel off-grid hybrid power system comprised of solar photovoltaic, wind, and hydro energy sources. *Applied Energy*, 133, pages 236–242. https://www.sciencedirect.com/science/article/pii/S0306261914007156, accessed 4 April 2018.

30. Heydari, A. and Askarzadeh, A. (2016, March 1). Optimization of a biomass-based photovoltaic power plant for an off-grid application subject to loss of power supply probability concept. *Applied Energy*, 165, pages 601–611. https://www.sciencedirect.com/science/article/pii/S0306261915016682, accessed 4 April 2018.
31. Salas, V., Suponthana, W., and Salas, R. (2015, November 1). Overview of the off-grid photovoltaic diesel batteries systems with AC loads. *Applied Energy*, 157, pages 195–216. https://www.sciencedirect.com/science/article/pii/S0306261915009149, accessed 4 April 2018.
32. IEEE Standard 1547-2018 (Revision of IEEE Standard 1547-2003). IEEE standard for interconnection and interoperability of distributed energy resources with associated microgrid system serving a group of customers.
33. Cordiner, S., Mulone, V., Giordani, A., Savino, M., and Jensen, J. (2015, April 15). Fuel cell-based hybrid renewable energy systems for off-grid telecom stations: data analysis from on field demonstration tests. *Applied Energy*, 192, pages 508–518. https://www.sciencedirect.com/science/article/pii/S0306261916312715, accessed 4 April 2018.
34. Qoaider, L. and Steinbrecht, D. (2010, February). Photovoltaic systems: a cost competitive option to supply energy to off-grid agricultural communities in arid regions. *Applied Energy*, 87(2), pages 427–435. https://www.sciencedirect.com/science/article/pii/S0306261909002529, accessed 4 April 2018.
35. Ghatikar, G., Mashayekh, S., Stadler, M., Yin, R., and Liu, Z. (2016, April 1). Distributed energy systems integration and demand optimization for autonomous operations and electric grid transactions. *Applied Energy*, 167, pages 432–448. https://www.sciencedirect.com/science/article/pii/S0306261915013537, accessed 4 April 2018.
36. Manchester, S., Swan, L., and Groulx, D. (2015, January 1). Regenerative air energy storage for remote wind–diesel micro-grid communities. *Applied Energy*, 137, pages 490–500. https://www.sciencedirect.com/science/article/pii/S0306261914006527, accessed 4 April, 2018.
37. Han, W., Chen, Q., Lin, R., and Jin, H. (2015, January 15). Assessment of off-design performance of a small-scale combined cooling and power system using an alternative operating strategy for gas turbine. *Applied Energy*, 138, pages 160–168. https://www.sciencedirect.com/science/article/pii/S0306261914011040, accessed 9 January 2020.

12 Energy Blockchain—Advancing DERs in Developing Countries

Alain G. Aoun

INTRODUCTION

Blockchain, a database shared across a network of computers, is a secure peer-to-peer trading platform that has gained the interest of the world's new businesses and industries. In a future in which consumers are encouraged to become producers, and everyone is encouraged to become entrepreneurs, energy blockchains have an important role in smart grid frameworks. Blockchain has the potential to change people's digital lives and the ways transactions are processed. By eliminating the roles of third parties in transactions, blockchain can make our systems more efficient, thus reducing cost and improving reliability. As the renewable energy sector continues to develop, the energy market is shifting toward smart decentralization which creates new markets for microgrids. Nevertheless, in developing countries, the deployment of distributed energy generation faces economic and financial barriers. Therefore, in countries lacking distributed energy resources (DER), the integration of blockchain and peer-to-peer energy trading creates incentives for end-users to invest in DERs and to develop microgrids. When blockchain is integrated in the power network, it enables the tracking of energy generated by DERs, supports the reporting of savings from energy conservation measures, and facilitates the trading of CO_2 credits.

How can the blockchain be modified to serve the energy market and create opportunities for energy producers and consumers? This chapter addresses this question by providing an overview of the blockchain mode of operation, highlighting the potential of the blockchain in the energy trading business, and identifying the challenges and barriers confronting the development of the energy blockchain.

The *World Energy Issues Monitor* published a report by the World Energy Council that offered an annual survey of key challenges and opportunities faced by energy leaders in managing robust energy transitions [1]. Blockchain was identified as one of the most critical uncertainties within the digitalization elements and was perceived by global energy leaders to be an issue of relatively high impact and uncertainty [1]. Blockchain has the ability to change the way we manage, organize, record, and verify transactions. According to Don Tapscott, Tapscott Group, the blockchain technology that underpins cryptocurrency could revolutionize the world economy. Blockchain's model provides an opportunity to shift from a centralized transaction structure (exchanges, trading platforms, energy companies) to a decentralized system based on peer-to-peer (P2P) trading.

The world's electric grids are undergoing transformations to apply intelligent technologies as a result of policies that encourage the use of renewable and DERs. Such policies increase the involvement of electricity consumers in managing and producing energy. Improving the reliability of today's energy supply system is hindered by the existing infrastructure of the centralized power supply system called the electric grid. The grid's unidirectional centralized energy system that supplies electricity to passive customers using traditional sources of energy is shifting to a dynamic, bi-directional network which is increasingly dependent upon DERs (e.g., intermittent renewable energy, electric vehicles, and energy storage systems). The diversity of the technologies, sources of energy, and power purchase agreements (net metering and feed-in tariffs) have important roles in advancing the use of DERs. In underdeveloped countries, DERs face economic, social, and environmental challenges and barriers. These include community culture, dissatisfaction with the existing infrastructure, scant financial incentives, and lack of enabling legislation.

In communities that have low DER penetration and a high demand for clean energy, the integration of blockchain and P2P energy markets can be an incentive for community residents to develop DERs.

In this chapter, we highlight the key challenges and barriers facing the implementation of DERs in developing countries. We also identify how the application of blockchain in the energy sector can introduce new incentives to develop DERs in developing countries based on worldwide business startups.

BLOCKCHAIN OPPORTUNITIES FOR ENERGY TRADING

A blockchain is a distributed chronological ledger that is hosted, updated, and validated by several peer nodes, rather than by a single centralized authority. Eliminating the central authority and having immutable transaction records that are validated by several peers, the blockchain increases the simplicity, speed, and transparency of transactions between two peers. An example of the implementation of blockchain is the cryptocurrency Bitcoin. While credit card transactions require validation from a bank and require time, blockchain doesn't require central validation and two-party transactions happen immediately [2].

Since blockchain offers a secure peer-to-peer trading platform, it has gained the interest of businesses and industries in energy markets. The blockchain promises a transactional platform that is fast and highly secure, has lower incidents of error, and reduces capital requirements [3]. It allows companies to automate more while processing greater volumes of data with fewer people at lower cost and risk. Companies that market energy are dealing with greater requirements for reporting, transparency, and security of data, which increases energy trading process costs and the need for personnel and resources. Energy blockchain provides solutions for these challenges. The energy sector is one of the industries where blockchain can have substantial impact.

Traditional transactional models are based on a centralized ledger structure (See Figure 12.1). Transactions between network nodes occur only through an intermediary third party who maintains the ledgers. Involvement of an intermediary is often necessary because it creates trust when transaction partners are unacquainted. Intermediaries usually charge fees for their services. The involvement of intermediaries increases the processing time required for transactions. Since all transactions are linked and stored on a central server or infrastructure, centralized structures have the disadvantage of a single point of failure. Alternatively, decentralized systems, such as the one offered by blockchain, have different network nodes that can interact directly with each another without an intermediary. There are many ways that transaction ledgers can be maintained. With a secure P2P distributed ledger technology, problems associated with centralized structures can be resolved.

In a peer-to-peer energy trading system operating on a blockchain platform, any individual member of the network can enter into an energy exchange with any other network member without restrictions from a centralized authority or the need for an intervening third party. P2P energy exchanges are possible using smart contracts in which energy transactions are immediate, automated, and flexible based on supply and demand in the system. Smart contracts are an appealing feature of blockchain technology. A *smart contract* is an executable code that operates on the blockchain to facilitate, execute, and enforce an agreement between untrusted parties without the involvement of a trusted third party [4]. The purpose of a smart contract is to automatically execute the terms of an agreement once its specified conditions are validated.

Blockchain technology has the potential to transform the entire contractual lifecycle by minimizing the need for human intervention from the time of trade execution to the time of payment. Blockchain technologies will improve market efficiency and potentially open and disrupt energy markets creating unforeseen opportunities. Among these is providing incentives for DER development in developing countries.

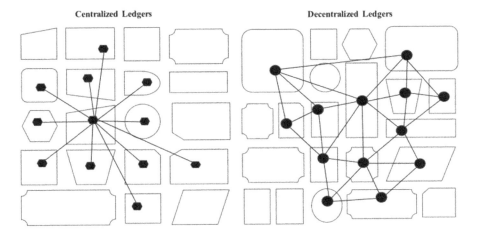

FIGURE 12.1 Centralized ledgers versus decentralized ledgers.

DER CHALLENGES IN DEVELOPING COUNTRIES

Though microgrids provide a platform to implement more cost-effective DERs, their development faces numerous technical challenges. Infrastructure associated with grid capacity and stability are barriers to the integration of renewable energy (RE) resources and distributed power generation. A microgrid's need for dual-mode switching functionality between grid-connected and island mode remains technically challenging. The conversion to island operating mode has two forms: 1) executing a *black start*, which allows a short period of outage before reenergizing the microgrid in island mode; and 2) performing a seamless transition within a short time period after disconnecting from the main grid [5]. The transition of reconnecting to the main grid also creates technical issues. Resynchronizing the two grids is sometimes sensitive. Like any power system, microgrids need protection schemes against external and internal faults.

The lack of skilled utility workers in some regional markets limits the implementation and development of DERs. Another barrier is the lack of awareness and knowledge of personnel about the availability and performance of renewable energy.

The economic challenges that hinder the development of microgrids which use renewable energy sources include market and financial barriers. Identified market barriers include the following:

- Inconsistent pricing structures can disadvantage renewables, especially those in long-term power purchase agreements.
- Subsidies for fossil fuels used to generate electricity make it difficult for RE resources to compete.
- The failure of pricing methods to include externalities such as health, social, and environmental costs underprices electricity that is generated using fossil fuels.

Financial barriers include lack of adequate funding and financing for renewable energy. The commercialization of DERs heavily depend on reducing the production costs of RE, storage technologies, and energy management systems.

Many developed countries and developing countries are adopting and promoting renewable energy resources and distributed generation to reduce their dependency on fossil fuels. Increasing the penetration of RE into the power generation system enhances their energy security. These countries have established numerous policies, directives, and standards, and set targets to support the implementation of distributed generation, RE, and microgrid development. The problem is not the absence of regulation but its weaknesses and ambiguities. Some regulations are undefined or not clearly stated. An example is Article 16(1) of the European Union's 2009/28/EC Renewable Energy Directive which requires member states to take *appropriate steps* to develop transmission and distribution grid infrastructure to allow the *secure operation of the electricity system*. These directives can be diversely and subjectively interpreted.

Another barrier to the development of RE resources is the adequacy of existing transmission lines to transmit energy from the points of generation to the points of consumption. Existing regulatory barriers hinder efforts to deploy renewables and expand microgrid markets. These include:

- lengthy administrative procedures for approval and permit
- policy instability with sudden changes and stop-and-go situations
- price competitiveness and cost friction, such as the direct and indirect costs related to the process such as commissions, interest rates, taxes, etc.

Identifying renewable energy zones is a regulatory barrier to the development of RE resources. Site selection is a regulatory issue during the design phase of large-scale RE power generation projects due to possible site conflicts with wildlife habitats, water reserves, and other environmental concerns. Without clearly defined policies and regulatory instruments associated with distributed generation grid penetration, it is improbable that these systems can survive.

INCENTIVES FOR DEVELOPING COUNTRIES

In industrialized countries, blockchain applications compete with advanced technological solutions supported by solid infrastructure and are embedded in the frameworks of trusted public institutions that enforce local regulations. In developing countries, this framework is not always available and often basic services (e.g., access to bank accounts) may be inaccessible. Alternatively, the market penetration of smartphones in many developing countries is as high as in industrialized countries. This has led many developing countries to adopt blockchain-based mobile money transfer applications. An example is M-Pesa, a mobile phone-based money transfer, financing, and micro-financing service launched in 2007 by Vodafone. According to Vodafone, after ten years of operation M-Pesa is now available in ten countries with more than 29.5 million active customers averaging 614 million monthly transactions [6]. The M-Pesa case is one of several examples that proves that developing countries offer fertile markets in which to implement new financial platforms.

The use of blockchain in the energy sector offers another potentially successful application. Next, some of the new ideas for blockchain applications in the energy sector based on real-world examples are introduced.

BLOCKCHAIN AND SMART METERS

Using blockchain and cryptocurrencies for monetary transactions is an achievable and tested application of blockchain technology in the energy sector. Blockchain eliminates the need for a third party in money transfer transactions and reduces transfer costs by eliminating administrative fees, bank fees, and commissions. An example of such an application is the United Nations World Food Program (WFP). The WFP is expanding its Ethereum-based blockchain payments system to avoid the transfer fees incurred when using the conventional banking system. The WFP performed a pilot test in the Jordanian refugee camp of Azraq to successfully facilitate cash transfers for over 10,000 refugees on its blockchain payments platform. According to Bernhard Kowatsch, the chief of Munich WFP's innovation laboratory, the pilot project saved the agency $150,000 a month and eliminated 98% of the bank-related transfer fees [7]. Blockchain decreases fundraising and operational costs, improving transparency, accountability, and control over how funds are used. Savings from using the blockchain's platform can be reinvested.

An early application of the blockchain in the energy sector is through the installation of smart pre-paid electric meters that provide power to the end-user once money is transferred from the end-user's account to the utility account. This application improves end-user payment discipline and reduces the cost of reading meters, billing, and collections. In Lebanon and many other countries, utility companies face difficulties in collecting amounts due on unpaid invoices [8]. Pre-paid electric meters offer a solution which decreases utility company losses, leads to lower energy rates, and avoids increases in electricity prices.

Another application of blockchain for energy sector money transfers is demonstrated by the South African company Bankymoon who use Bitcoin to perform remote payment transactions with compatible smart meters. The application works as follows: assume a donor wants to support a school in a developing country; the donor can send cryptocurrency directly to the school's smart meter, enabling electricity from the grid to be supplied to the school.

SMART CONTRACTS AND P2P ENERGY TRADING

Decentralized cryptocurrencies have been much more successful than many prior incarnations of electronic cash [9]. The ability to conduct smart contracts via a platform called Ethereum has increased blockchain applications. Ethereum is a public blockchain platform that can support advanced and customized smart contracts using a Turing-complete programming language [10,11]. The Ethereum platform supports withdrawal limits, loops, financial contracts, and gambling markets. NXT is another public blockchain platform that allows non-customized smart contract templates. Smart contracts have a potentially major role in the integration of the blockchain in the energy sector.

One application of energy smart contracts is managing peer-to-peer energy transactions in private community microgrids (see Figure 12.2). U.S.-based startup TransActive Grid, launched in early 2016, enables its users to trade energy using smart contracts via blockchain. The project consists of connecting five Brooklyn, New York households with solar photovoltaic (PV) systems with nearby consumers interested in buying excess electricity produced by their neighbors [12]. Similar initiatives have been launched by Power Ledger in Perth, Australia and Grid Singularity in Austria. In addition to energy exchange, Grid Singularity offers data analysis, benchmarking, smart grid management, green certificate trading, and energy trade validation. Collected technical and financial data can be used for real-time asset valuation of DER power plants. The collected data can be beneficial for refinancing or selling a power plant allowing investors to perform online due diligence. Other uses of the collected data include assessments of generation capacity, power availability, pricing, forecasting, and energy trading [13]. For electric utilities, collected data has a future role in the development of grid-balancing mechanisms.

Energy Blockchain—Advancing DERs   195

FIGURE 12.2 Peer-to-peer energy trading in community microgrid.

Using blockchain platforms for P2P energy trading in community microgrids provides a competitive advantage for innovative retailers and a better return for excess generation, plus a low-cost, transparent, secure, and instant payment system for electricity transactions. This model is ideal for households, building owners, and retailers who produce renewable electricity using DERs.

Energy Backed Currencies

The lack of financial incentives or funds is the primary barrier facing the development of distributed energy resources. This challenge applies to both developed and developing countries. In 2016, approximately a third of Germany's electricity consumption was generated using renewables. During one winter day in February 2020, over 75% of the country's electricity was generated by wind power. Despite the success of this transition to renewables, the central government recently reduced financial incentives for new RE installations, particularly solar PV [13].

An energy-backed currency might be the type of incentive needed to improve the economic feasibility of renewable energy-based power generation. The concept of an energy-backed currency is similar to the gold reserves that are used to stabilize national currencies. An application of this concept is offered by the SolarCoin Foundation, a nonprofit organization. They focus on promoting solar energy generation by providing SolarCoin to solar energy producers [14]. Nick Gogerty and Joseph Zitoli, two founders of the SolarCoin Foundation, had an idea for an energy-backed currency called DeKo in 2011 [15]. In 2014, the DeKo was transformed by the SolarCoin Foundation into a digital asset reward program for RE installations based on a new cryptocurrency. SolarCoin is a digital currency backed by solar-generated electricity, electronically representing a verified 1 MWh of solar-generated electricity. The main purpose of this initiative was to provide an incentive to produce solar energy by rewarding SolarCoin to producers of solar electricity. SolarCoin is active in 17 countries and is intended for worldwide circulation. It can be exchanged for other cryptocurrencies or conventional currencies. Holders can use their SolarCoin to pay for products and services from participating merchants and service providers.

The *Jouliette* is another energy-backed currency named after the joule, an SI energy measurement unit. The Jouliette is backed by physical energy production. In September 2017, Spectral and Alliander launched the Jouliette token at De Ceuvel, a city playground for innovation and creativity in Amsterdam and a showcase for sustainable urban development [16]. A private smart grid made the Jouliette model feasible for the De Ceuvel community. With the Jouliette token, the De Ceuvel community members make P2P energy transactions using blockchain. This ensures that the transactions are secure, without the need of a bank or trusted third party. Transaction histories can be shared with all community members enabling automatic verification. The De Ceuvel community is exploring further applications for the Jouliette tokens such as using it to trade for goods and services within the community [17].

Energy-backed currencies using blockchain might provide incentives for communities in developing countries, especially if combined with smart contracts and P2P energy trading within community microgrids.

BLOCKCHAIN FOR ELECTRIC VEHICLE CHARGING

In some countries the private sector is not allowed to supply electricity to the utility grid, or there are severe limitations which makes this infeasible. When incentives and development funds for DERs in developing countries are lacking, using blockchain for charging plug-in electric vehicles is an alternative. Widespread use of electric vehicles (EV) becomes more feasible when drivers can universally access charging stations [18].

A concern is how to simplify billing at charging stations, especially ones located in public parking spaces. Blockchain can be used to build a simple, secure billing model for EV battery charging and thus help remove this barrier. EV drivers could park their vehicle at any blockchain-compatible charging station while the vehicle autonomously logs on to a charging station and recharges automatically. Once the vehicle leaves the parking space, the charging station automatically invoices the driver for the electricity received using blockchain technology. Since EVs have energy storage systems, excess energy stored in their batteries can be fed back to the network with the owner earning money for storage services. The DAV Foundation's blockchain-based application for EVs is an example of this application which creates possibilities for vehicle owners, mobility services, riders, and shippers. The DAV Foundation's token enables exchange of energy between the grid and EVs. The platform records transactions and provides compensation using DAV tokens [19].

BlockCharge is another newly established blockchain application for EV battery charging. It uses the Ethereum blockchain to facilitate EV charging. This project was launched by the German utility Innogy, assisted by a company named Slock.it. The concept of BlockCharge is based on the *smart plug*, which can be used as a normal electrical outlet plug but is linked with an identification code [20]. Users install an application on their smartphones to authorize the EV charging process. It connects

Energy Blockchain—Advancing DERs

to Ethereum blockchain which negotiates the price, records the charging data, and manages the payment process (see Figure 12.3) [20]. BlockCharge uses a business model based on the one-time purchase of the smart plug and a transaction fee for the charging process [20]. BlockCharge is aiming for a worldwide authentication, charging, and billing system with no intermediary. Once induction charging for electric vehicles becomes more widely adopted, applications like BlockCharge will manage the entire EV charging process.

The application of blockchain technology for charging electric vehicles represents a change in how private businesses interface with their customers. This model (see Figure 12.3) is ideal for small businesses, shopping malls, office buildings, and car parks that offer customers renewable energy charging for their plug-in EVs.

KEY CHALLENGES AND BARRIERS

Blockchain energy-based applications provide solutions to the challenges and barriers associated with the development of DERs. Despite the number of blockchain startups that successfully operate worldwide, there are technological constraints that challenge the broader implementation of blockchain in the energy sector.

Unlike cryptocurrency transactions, energy transactions involve physically delivering the electricity. A blockchain-based energy model must reflect the physical configuration of the power grid. The large scale of the blockchain required might be a challenge due to the multiple interferences with the distribution or transmission networks. Implementing large-scale energy blockchains might cause network capacity problems. Blockchains such as Ethereum and Bitcoin are only capable of processing three to five transactions per second [22]. Therefore, transaction capacity has to increase significantly to counter this challenge. It is technically easier to implement a blockchain-based energy model in small communities or in isolated or independent microgrids.

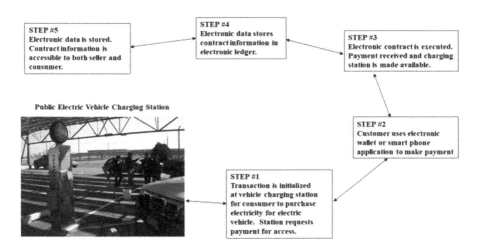

FIGURE 12.3 Blockchain-based electric vehicle charging [21].

Another problem is encountered with the installation of blockchain-compatible smart meters. In addition to cost, technical challenges arise from the physical connection between the electrical power system and the blockchain. These include information sharing, privacy, and security.

Information sharing is an issue because of the transparent nature of the blockchain ledger. Due to this, it might be possible to deduce from the blocks information related to energy transactions such as consumption patterns, and the quantity and price of purchased or sold energy. Such information might give indications to competitors and be a problem for utility providers and industrial consumers.

Privacy is a concern in a fully functional blockchain-operated electrical energy market. Because of the transparency of the ledger of the blockchain-trading infrastructure it might be possible to deduce information about closed electrical energy transactions. This could compromise the privacy of consumers, electrical utilities, and other energy suppliers within the electrical power system.

To be considered as a large-scale energy solution, blockchain has to prove that it is more effective and secure than other available alternatives. Due to the importance of energy in national security plans, the security of the energy supply and data should outweigh other considerations. Stability of a digital energy system is crucial to its successful operation. It must operate without internal complications and be shielded against cybercrimes and espionage. Thus, cybersecurity is a major challenge that must be resolved prior to the broader implementation of blockchain in the energy sector.

CONCLUSION

The fourth industrial revolution is associated with a global trend toward a decentralized energy grid. Blockchain is the technology that can move the energy system from its centralized form to a smart decentralized network more appropriate for microgrid deployment. Blockchain is a database shared with a network of computers that acts as a record of transactions [23]. Since transaction records are stored on multiple computers and updated simultaneously, it's much more secure than a centralized system [23]. Blockchain can be viewed as an important overlay on the internet similar to the world wide web in the 1990s.

Nevertheless, blockchain market infiltration will be met with resistance because it represents an extreme change to the present ways of doing business, especially since it eliminates the need for trusted third party intermediaries. The first stage of phasing out intermediaries was initiated with the internet and blockchain will be its second stage. Yet this resistance might be less in developing countries than in industrialized ones. Like the integration of smartphones, developing countries represent fertile markets for the growth and prosperity of blockchain technology. Blockchain is an ideal technology for disrupting energy transaction processes.

The ability to conduct smart contracts via a platform is shown to enable and expedite peer-to-peer energy trades which will further decentralize the energy markets of the future. Electric vehicle charging offers an example. Considering the blockchain potential in developing countries from the distributed energy perspective, this chapter identifies the challenges and barriers associated with the development of DERs in developing countries and highlights opportunities associated with the integration of blockchain-based solutions to overcome these challenges. Blockchain is offered as an example of a transaction system that can support and incentivize microgrid development. Moreover, this chapter provides a review of the existing worldwide energy blockchain startups and details the major confronted limitations, related to scalability, security, stability, and privacy.

ACKNOWLEDGMENTS

This chapter is a revised version of a peer-reviewed article entitled "Energy Blockchain—Opportunities and Challenges in Advancing Distributed Energy Resources" that was originally published in the *International Journal of Strategic Energy and Environment Planning*, Volume 1, Issue 4, Stephen A. Roosa, Ph.D. (editor) in November 2019. The editor extends thanks and credit to the Association of Energy Engineers, Atlanta, Georgia for use of previously published material and to the author of the chapter who approved that the article be republished.

REFERENCES

1. World Energy Council. The developing role of blockchain. White paper in collaboration with Pricewaterhouse Coopers. https://www.worldenergy.org/assets/downloads/Full-White-paper_the-developing-role-of-blockchain.pdf, accessed 20 May 2020.
2. Thakkar, A. *How Blockchain and Peer-to-Peer Energy Markets Could Make Distributed Energy Resources More Attractive*. Duke University: Durham, North Carolina.
3. Deloitte LLP (2016). Blockchain application in energy trading. https://www2.deloitte.com/content/dam/Deloitte/global/Documents/Energy-and-Resources/gx-Blockchain-applications-in-energy-trading.pdf, accessed 20 May 2020.
4. Maher, A. and Moorsel, A. (2017). Blockchain based smart contracts: a systematic mapping study. *Fourth International Conference on Computer Science and Information Technology*, 125–140.
5. Mariya, S., Wina, C., Josep, G., and Vasquez, J. (2014). Microgrids: experiences, barriers and success factors. *Renewable and Sustainable Energy Reviews*, 40, pages 659–672.
6. Vodafone (2017). Vodafone endorses the International Day of Family Remittances. https://www.vodafone.com/content/index/what/m-pesa.html, accessed 16 November 2018.
7. Das, S. (2018, February 19). Banks begone: UN's world food program builds on Ethereum blockchain money transfers. https://www.ccn.com/banks-begone-uns-world-food-programme-builds-ethereum-Blockchain-money-transfers, accessed 17 November 2018.
8. World Bank (2018, January 31). Report No. 41421-LB, Republic of Lebanon electricity sector public expenditure review. Sustainable Development Department Middle East and North Africa Region.
9. Delmolino, K., Arnett, M., Kosba, A., Miller, A., and Shi, E. (2016). Step-by-step towards creating a safe smart contract: lessons and insights from a cryptocurrency lab. *International Conference on Financial Cryptography and Data Security*, pages 79–94.
10. Buterin, V. (2019). A next-generation smart contract and decentralized application platform. https://github.com/ethereum/wiki/wiki/White-Paper, accessed 23 February 2020.

11. Wood, G. (2014). Ethereum: a secure decentralized generalized transaction ledger. *Ethereum Project Yellow Paper.*
12. Rutkin, A. (2018). Blockchain-based microgrid gives power to consumers in New York. https://www.newscientist.com/article/2079334-Blockchain-based-microgrid-gives-power-to-consumers-in-new-york, accessed 23 February 2020.
13. Burger, C. (2017, December 15). Blockchain and smart contracts: pioneers of the energy frontier. https://www.ibtimes.co.uk/Blockchain-smart-contracts-pioneers-energy-frontier-1651650, accessed 23 February 2020.
14. Bloomberg (2018). Company profiles – SolarCoin Foundation. https://www.bloomberg.com/profiles/companies/1627999D:BZ-solarcoin-foundation, accessed 16 November 2018.
15. Gogerty, N. and Zitoli, J. (2011, January). eKo: an electricity-backed currency proposal. *SSRN Electronic Journal.* doi: 10.2139/ssrn.1802166.
16. Jouliette. https://jouliette.net/index.html.
17. Kastelein, R. (2017, September 27). Spectral and Alliander launch Blockchain-based renewable energy sharing token. https://www.the-blockchain.com/2017/09/27/spectral-alliander-launch-blockchain-based-renewable-energy-sharing-token, accessed 23 February 2020.
18. Pricewaterhouse Coopers (2018). Power and utilities. Blockchain – an opportunity for energy producers and consumers? Study conducted by PwC on behalf of Verbraucherzentrale (consumer advice center) NRW, Düsseldorf, Germany. www.pwc.com/utilities.
19. Linnewiel, R. and Berman B., editors (2018, July 5). With blockchain technology, electric vehicles become energy-storage devices on wheels. https://medium.com/davnetwork/with-blockchain-technology-electric-vehicles-become-energy-storage-devices-on-wheels-9e9659e32f, accessed 23 February 2020.
20. Futures Centre (2017, September 29). Prototype blockchain electric vehicle charging and billing system. https://medium.com/signals-of-change/prototype-blockchain-electric-vehicle-charging-and-billing-system-cf8998fc31e8, accessed 17 November 2018.
21. Stocker, C., Jonghe, D., and Ruther, M. (2018, January 18). Blockchain 2.0: exchanging value in the physical and digital world, continuously and simultaneously. *International Business Times.* https://www.ibtimes.co.uk/blockchain-2-0-exchanging-value-physical-digital-world-continuously-simultaneously-1656257, accessed 23 November 2018.
22. Young, J. (2018, June 3). Vitalik Buterin: Ethereum will eventually achieve 1 million transactions per second. https://www.ccn.com/vitalik-buterin-ethereum-will-eventually-achieve-1-million-transactions-per-second, accessed 23 February 2020.
23. Schutte, S. (2018). The power next door. *Centrica.* https://www.centrica.com/platform/the-power-plant-next-door?fbclid=IwAR2jw2HGv_XVbP0GRP7IFddsjLTPAodhHsurQXjzaPdFWlTntsvBJWtkCBI, accessed 4 December 2018.

13 Smart Microgrids

INTRODUCTION

This chapter considers smart microgrids. It explains what they are, how they work, and the key differences between traditional and smart microgrids. The main objective of this chapter is to discuss how smart systems are integrated into microgrids. Examples include advanced energy technologies and systems that enable the efficient delivery of electrical energy.

Microgrids can generate electricity using fossil fuel, renewables, and other clean energy technologies. Clean energy-based direct current (DC) microgrids are considered to be a revolutionary power solution [1]. Matching the electrical energy supplied by microgrids with loads can be difficult. The intermittency of power generation associated with renewable energy (RE) can result in problems with power generation, distribution, and demand which contribute to electric grid instability [1]. Smart microgrids use sensing systems and direct digital control (DDC) systems to overcome these issues and enable their customers to connect to the supply networks in real time.

Smart grids are modern electricity distribution systems that monitor, protect, and automatically optimize the operation of interconnected elements including generation equipment, high-voltage distribution, automation systems, and energy storage [2]. They are characterized by a bi-directional flow of electricity and information to create an automated, widely distributed energy delivery network [2]. Like microgrids, smart microgrids can generate AC or DC electricity. They offer the benefits of distributed computing and communications to deliver real-time information and enable instantaneous balancing of electrical supply and demand at the level required for each discrete device. A control system is a key feature of smart grids. Smart grid controls are typically the central computer and load dispatch location from which the distribution of total system generation is managed, and the monitoring and control of system stability and load frequency is performed. Algorithmic software platforms provide decision-making and predictive intelligence capabilities. Some elements of the control system are decentralized and capable of independent operation.

SMART MICROGRIDS AND THEIR BENEFITS

As microgrids are being connected to smart grids, new theories have emerged redefining how microgrids are deployed. These nascent microgrids have variously been called advanced microgrids, advanced remote microgrids, or smart microgrids. There are many benefits associated with smart microgrids and they employ a wide range of technologies.

Advanced Microgrids

The Smart Energy Power Alliance (SEPA) has a working group that advocates emerging concepts associated with microgrids [3]. SEPA has proposed a definition for *advanced microgrids* as "electricity delivery networks that are intelligently managed, energy and resource efficient systems" [3]. They "interconnect, interoperate, and optimize the performance of loads, distributed resources, and energy storage using a layered control scheme, within defined electrical boundaries that act as a single controllable entity with respect to the macrogrid at the point of common coupling… they can island, disconnecting from the grid to enable it to operate in both grid-connected or island modes" [3]. Advanced microgrids "balance supply with demand in real time, schedule dispatch of resources, and preserve grid reliability" [3]. They are designed to optimize energy profiles and efficiencies, enhance island performance, and maximize the economic benefits associated with grid interconnectivity [4].

A report by Sandia National Laboratories notes "a major goal for advanced microgrid systems is to develop promising new solutions to integrating advanced microgrids capable of operating in parallel with the utility distribution system and transitioning seamlessly to an autonomous power system complete with its controls, protection and operating algorithms. It is expected that advanced microgrids will be fielded in a wide variety of electrical environments ranging from substations to building-integrated systems" [4]. It also suggests that applying new components to advanced microgrids will entail advanced controls, operational methodologies and protocols, and secure communications [4].

Advanced Remote Microgrids

A subcategory of advanced microgrids is *advanced remote microgrids*. These microgrids operate in very isolated locations, often in harsh environments such as in caves, deserts, artic areas, or places at high elevations. They are often required to provide continuous power and may require sophisticated and hardened control systems. While the definition of advanced remote microgrids is somewhat unclear, they have identifiable attributes. Holdmann and Asmus have compiled some of their unique features which they categorize as either mandatory or preferred [3].

Mandatory Microgrid Features
- High penetrations of local distributed and local renewable energy resources. Remote microgrids should be capable of achieving instantaneous penetration from RE generation of 100% of the load with an annual average penetration of 50% or more.
- At least one non-firm RE resource should be incorporated into the system. Managing high penetration levels of variable RE resources such as wind or solar is much more complicated than achieving similar levels of penetration from baseload RE sources such as hydropower or geothermal [3].

Preferred Microgrid Features

- If the system uses a diesel generator, it must be capable of operating with the diesel generator turned off when adequate RE resources are available.
- The controls platform will usually feature state-of-the-art technology capable of managing high penetrations of RE resources. This control platform would lean toward a distributed grid edge intelligence, with auxiliary components required to regulate voltage and frequency, and leverage smart inverters or direct current modular technologies to achieve continuous reliability and resiliency.
- Smart meters or some other form of advanced telemetry are used for utility billing purposes. This capability enables individual energy costs to be allocated according to actual energy consumption. (This is vital for access to energy from remote microgrids in places that operate using mobile phone-enabled, pay-as-you-go business models.)
- The system will likely incorporate some form of energy storage with hybrid energy storage solutions being ideal. This might include batteries, flywheels, ultra-capacitors, thermal storage, or pumped hydro storage.
- The advanced remote microgrid would not be limited to the provision of electricity but might also provide both heating and cooling (trigeneration) and ultimately serve as a platform for other essential services.
- While capable of integrating the latest technology advances in DER, energy storage, and controls, the design would reconcile operations and maintenance challenges, and leverage remote monitoring, training, and troubleshooting opportunities [3].

Examples of advanced remote microgrids include the ones serving Kongiganak, Alaska or McMurdo Station in Antarctica.

DIFFERENCES BETWEEN SMART GRIDS AND SMART MICROGRIDS

The key difference between a *smart grid* and a *smart microgrid* is the scale of the network. The configuration of smart microgrids in some ways replicates the architecture of regional smart grids but on a smaller, more local scale. *Smart grids* are utility-scale systems with substantial and large transmission and distribution systems that operate at high voltages. *Smart microgrids* are local systems that use distributed energy resources, and have internal transmission, smaller loads, and lower voltages. Smart microgrid, therefore, refers to more than simply having a smart electric meter.

Smart grids and smart microgrids are both characterized by substantial use of digital monitoring and control systems with automated decision-making capabilities. Smart microgrids are often considered to be a subset of the smart grid when the microgrid is operating in grid-connected mode. When operating in island mode, the smart microgrid is independent and not considered to be a subset of the grid. Innovative smart grid and microgrid configurations use three primary layers of technologies [5]:

Power layer: consists of the conventional electric value chain, consisting of generation, transmission, and distribution.

Communications layer: those technologies, such as home, local, fluid, and wide area networks, that connect and enable data transmission from the power layer with the applications layer.

Applications layer: the higher-level layer of software and sensors that provide control and vigilance mechanisms for monitoring utility distribution and protecting the power system. It also provides electronic data storage and software [5].

Microgrids when interconnected can supply power to consumers using either grid power, microgrid generated power, or a combination of both. Smart microgrids can operate in autonomous mode and provide power for uninterruptable power supply (UPS) systems when main grid power is disrupted [1]. They can be configured as AC or DC networks or use a combination of the two. Any sort of electrical generation technology can be used. Smart microgrids might employ distributed automation capabilities that allow individual devices to sense aspects of grid operating conditions and make adjustments to improve power flow and optimize performance [5]. An example is offered by Strunz et al. who proposed an application for smart DC microgrids using both solar and wind energy [6].

There are various other conceptual definitions for smart microgrids. A *smart microgrid* has also been defined as a more modern electrical grid that uses information and communications technology to gather and act on data collected about the activities of suppliers and consumers to improve the efficiency, reliability, economics, and sustainability of the production and distribution of electricity [7]. Others view smart microgrids as smaller, more automated and robust independent grids (an extension to the regular microgrids) that use newer technologies, software, and intelligent controls to manage the flow of electricity within networks [8]. Their key components are distributed energy resources (DERs) such as solar PV or wind energy, electrical loads, and storage devices such as batteries [8]. The main features of smart microgrids are digital information and control, near-instantaneous response to optimize grid operations, and smart metering systems with real-time management capabilities for energy consumption and electricity storage systems [8].

BENEFITS OF SMART MICROGRIDS

The principle benefits of smart microgrids are associated with the technologies they deploy. Smart microgrids can be considered to be an upscaled version of microgrids. According to Niclas [8], the benefits of smart microgrids include:

Improvement over traditional grid: at the local level, smart microgrids are an efficient means of cost-effectively integrating consumers and networks with electricity distribution and generation.

Greater reliability: local power generation helps eliminate blackouts. Technologies such as sensors and smart switches are capable of anticipating and repairing power disturbances rather than using manual switches in traditional grids used to handle outages.

- *Cost-effective*: microgrids allow consumers to obtain electricity in real-time at lower costs, as they rely on locally generated power to avoid using costly peak electricity. The microgrid model also shields ratepayers from bearing infrastructure improvement costs that are common in a traditional grid model.
- *Encourages local employment*: small communities with microgrids create employment at the local level.
- A *methodology for future-proof grid*: conventional grids are costly to extend to local remote areas and may require many years to come online. Smart microgrids are much better positioned than the centralized grid to meet the future electricity needs of their customers. Local communities can reliably depend on them to meet their energy needs quickly and efficiently through small local generators, solar generation, wind turbine generators, etc.
- *Reduces carbon footprint*: smart microgrids are equipped to use waste heat from their locally generated electricity. They can reuse the energy that is produced during electricity generation for heating buildings or hot water and for cooling and refrigeration purposes [8].

It is clear that *smart microgrids* are an extension to regular microgrids and a new way of deploying microgrid technologies [8]. They use software and intelligent controls to manage electricity flows in networks [8]. They also use DDC systems to a greater extent and are remotely manageable.

Likely the key benefit of a smart microgrid has much to do with the costs. Since these types of microgrids often digitally optimize their operation and management, costs for electricity can decline over time. Electricity produced from RE resources supports the development of smart microgrids, using the electricity efficiently for various on-site applications [1]. Smart microgrids are highly automated and can operate independently with minimal human intervention.

TECHNOLOGIES USED FOR SMART MICROGRIDS

The technologies that are being deployed for smart microgrids involve enhancements to sensor arrays, actionable switching, and digital control systems. They are scaled to residential, commercial, and distribution system applications.

Sensor Systems

Sensor system architecture incorporates sensors that acquire information from the data environment, and uses communication devices and hubs to collect the sensor data and relay it to a central repository [2]. Intelligent electronic devices (IEDs) receive data from the sensors and power-generating equipment. Software scans the data acquired and focuses on information that is of diagnostic interest. Control actions are issued as needed such as tripping breakers or changing voltage levels to maintain desired conditions [2]. With today's micro processing technologies, IEDs can perform multiple protection, self-monitoring, communication, and control functions [2]. Sensors and software for smart microgrids with renewable generation (e.g., wind and solar) must enable real-time operation and often require dynamic stochastic optimal control (DSOC) with forecasting capabilities and characterization of power outputs [9]. The increased use of advanced sensors and high-speed communications networks on transmission systems is also improving the ability to monitor and control operations at substations and across transmission networks [10].

Advanced Metering Infrastructure

Metering systems for smart microgrids are more than electric meters that record KW and kWh usage. While smart microgrids use advanced metering infrastructure (AMI), they also deploy communication networks that transmit utility data at programmed intervals, and management systems that receive, store, and process the data obtained from the meters [10]. Electric usage data from AMI is highly granular and can also be sent directly to microgrid central controls, building energy management systems, customer information displays, and smart appliances [10]. AMI enables a wide range of capabilities that can provide significant operational and efficiency improvements to reduce costs [10], including:

- remote meter reading and remote connects and disconnects
- tamper detection to reduce electricity theft
- improved outage management from meters that alert utilities when customers lose power
- improved voltage management from meters that convey voltage levels along a distribution circuit
- measurement of two-way power flows for customers who have installed on-site generation such as rooftop photovoltaics (PV)
- improved utility billing and customer support operations [10]

Providing AMI with customer-based systems and real-time rate utility structures can be used to reduce electricity demand during peak periods [10]. For this to happen, interventions must be programmed to reduce demand at critical times. Peak demand reduction can lower costs by more than 30% depending on the rate design and type of customer system [11]. AMI enhances the operational efficiency of utilities and provides microgrid operators with information that helps them to more effectively manage their customers' energy use [10].

Technologies

The technologies used in smart microgrids are often custom applied and designed to resolve issues associated with their development and implementation. For example, one way to maintain instantaneous power balancing is to provide controls that actively resolve fluctuations in power and frequency in real time. Actively bringing multiple generation systems on- or off-line can pose problems. Smart microgrids employ advanced technologies and algorithms to resolve this. By using a decentralized architecture of multi-agent systems to perform microgrid control and load-balancing functions, all options for generation can be digitally viewed as hierarchically equal [9]. As there is no central agent, any can be removed or added without reconstructing the system; this capability improves system reliability, vulnerability, and flexibility [9].

Smart microgrid technologies are decentralized and highly granular. They begin with residential and small commercial applications and include those associated with the microgrid's electricity distribution system. They typically use renewable energy generation such as solar, wind, geothermal, or hydropower and may use battery storage systems (Figure 13.1). A summary of the categories of technologies used in smart microgrids is provided by the Galvin Electricity Initiative [12]:

Residential
- energy efficiency improvements that help consumers use less electricity and reduce electricity costs
- smart meters that enable bi-directional exchange of information (pricing, usage data, demand) and electricity
- programmable smart appliances and devices that energize when the price of electricity matches the end-user's desired price point
- user-friendly energy control systems that allow customers to interface with the smart microgrid to automatically control electricity usage [12]

Commercial
- advanced lighting technology with digitally programmable controls that are responsive to the cost of power, building occupancy, and locations of use
- heating and air conditioning control systems that continuously and automatically adjust building ventilation based on occupancy, air quality, electricity costs, or other factors
- nascent electricity generation technologies that provide electricity to individual buildings and are capable of conditionally supplying power to the host grid [12]

Electricity Distribution
- redundant design configurations that have secondary power sources when inclement weather conditions (high winds, storms, ice, etc.) interrupt power
- smart switches, relays, and sensors that replace less efficient equipment to allow the smart microgrid to manage and distribute electricity more efficiently and reliably
- accessible and protected infrastructure within structures (e.g., hardened or installed underground)
- DDC systems that continuously scan the data environment, identify and anticipate potential instabilities, and initiate corrective actions prior to disruption in electrical service [12]

EXAMPLES OF SMART MICROGRIDS

A defining component of smart microgrids is the digital control system that enables the data environment to be periodically polled, makes intelligent decisions based on

FIGURE 13.1 Grid-connected solar PV project near Amman, Jordon. (Courtesy of Mohammad Shehadeh, Philadelphia Solar.)

the information that is gathered, records a history of actions taken, and calculates the value of the results. The controls provide adjustment and fine-tuning when appropriate. While this process mimics human decision-making there are far more variables and conditions involved to be quickly managed by individuals. Smart microgrids are supported by their controller and defined as the resources—generation, storage, and loads—within boundaries that are directly managed by the controller and its capabilities [13]. The controller "manages the resources within the microgrid's boundaries, at the point of interconnection with the utility, interacting with the utility during normal operations" while defining "the microgrid's operational relationship with the distribution utility" [13]. It can be challenging to find microgrid controllers that fully utilize all customer-based resources [14]. DER controllers vary based on the types of generation. Renewable energy systems often require a custom controller that must be individually programmed based on the types of generation used by the microgrid. A solution is to construct a project-specific microgrid visualizer to enable monitoring of all operations [14]. There are many examples of smart microgrids under development or deployed.

Projects Under Development

Smart microgrids are revolutionizing our vision of supplying electricity to localities, energizing community involvement, and changing the way we view infrastructure for cities in the future. There are numerous examples of smart microgrids under development. The focus here is on models for future projects, their goals, and the keys to their success.

Jayant Kumar discussed the importance of developing smart microgrids in applications for sea ports and their facilities [15]. A new microgrid system has been suggested for Philadelphia Navy Yard Alstom [15]. The goals were to improve grid resilience, provide demand response, reduce emissions, and improve system efficiencies. At the level of the controller, the objectives were to provide islanding management, resynchronization and reconnection management, frequency and voltage management, microgrid protection, portfolio optimization, dispatch management, system simulation capabilities, and grid resiliency [15]. This is indeed a challenging menu of capabilities for a microgrid control system. The microgrid topology proposed was even more complex. It included the use of multiple microgrids with more than one utility interconnection. For example, one microgrid was based on power generated by a combine heat and power (CHP) plant combined with solar PV. Another was based on natural gas generation combined with solar PV, fuel cells, and energy storage. Its smart features included a microgrid controller system with field inputs and programming for islanding, voltage synchronization, and frequency management capabilities [15]. Unmanned and automated control hardware systems were provided for the substations.

Successful microgrid deployment involves modeling and analysis. Gerry et al. adopted an approach based on mathematical modeling for each component of a microgrid's distributed energy resources that includes generation (solar PV array, wind energy, hydropower and biogas-fueled generators) with battery backup [1,16]. They used HOMER (a software optimization tool) to determine the economic feasibility of DERs and ensure reliable power supply for load demand at minimal cost [1,7]. Mariam et al. performed a technical and economic analysis of micro generation for the development of community-based microgrid systems [1,16]. They performed this analysis for wind power-supplied micro generation systems for communities that lacked renewable energy feed-in-tariff policies [1,17].

Projects Deployed

An example of an operational smart microgrid is provided by the experience of rural Les Anglais, Haiti which uses a mesh network of wireless smart meters deployed at 52 buildings [18]. This microgrid is wireless and its architecture includes a cloud-based monitoring and control service plus a local embedded gateway infrastructure [18]. Each end-user's smart meter has a wireless radio that enables remote monitoring and control of the electrical services [18]. The smart meters communicate over a scalable multi-hop network back to a central gateway that manages system loads [18]. The gateway also provides an interface for an on-site operator and a cellular modem connection to a cloud-based backend that manages and stores billing and usage data [18]. The cloud enables occupants in each home to pre-pay for electricity at a particular peak power limit using a text messaging service [18]. The system activates each meter within seconds and locally enforces power limits with provisioning for theft detection [18].

Types of electrochemical storage used in smart grids are basically lead-acid, sodium-sulfur (NaS) batteries, and in some cases lithium-ion (LI) batteries [19]. Redox flow batteries also have potential as a microgrid storage solution because of their independent ratio of power and energy and cost-efficiency [19]. Panasonic Group (Sanyo) has deployed a smart microgrid with a large-scale storage battery system using LI batteries [19]. The system was installed at the Kasai factory in Japan in October 2010. The system charges the batteries with overnight electricity and surplus solar electricity and discharges it during the day [19]. The electricity energy storage (EES) system has more than 1,000 LI battery boxes, each box consisting of over 31 million cell batteries [19]. With the battery management system, the whole EES can be used as if it were just one battery. The capacity of the EES is approximately 1,500 kWh; the solar PV system can distribute 174,060 kW of DC power [19]. The system can reduce power by 15% during peak periods using the energy management controls.

Fort Bragg in North Carolina has developed a microgrid with smart components that is anchored by a renewable biogas plant and a 5 MW solar system [20]. With about 62 MW of backup generation at the base, the multi-phase project involved installing control systems to incrementally create a smart microgrid and integrating it with the post's distribution network, information technology, and communications infrastructure [12]. The $3.4 million project was among the first to link multiple backup power sources into a network using advanced controls and sophisticated management software [20]. Fort Bragg ultimately integrated a variety of distributed generation technologies to work in conjunction with the military base's utility infrastructure [12]. Since Fort Bragg owns its own electric distribution network the project enabled managers to monitor electrical generation from a central energy management location [12].

USE OF RENEWABLE ENERGY TECHNOLOGIES

For the development of smart microgrids, a strategic plan is important to match the proposed renewable power generation of the microgrid to that of the electricity demand of the consumers being served. This involves policy, technology, and organizational analysis. Microgrids can be a solution to the integration of different renewable energy systems.

Sustainable Smart Microgrids

A sustainable smart microgrid might consist of wind and/or solar, battery (energy storage), and thermal generation with a dynamically moderate load [9]. It is challenging to design an optimum smart microgrid with RE given storage costs, emissions considerations, and reliability issues [9]. Oversizing of resources, e.g., wind farm, solar farm, storage, and backup thermal, increases capital cost which is often unaffordable [9]. Undersizing microgrid resources creates system unreliability. Optimizing smart microgrid energy generation to achieve proper sizing of active resources to reduce costs while attaining a goal of zero net energy and emissions for the system is a multi-objective optimization problem with a large number of constraints [9].

Smart microgrids are characterized by the use of multiple sources of energy and extensive use of DDC systems with remote operation and monitoring capabilities.

Black & Veatch is headquartered in Overland Park, Kansas. The company constructed a microgrid to power its Innovation Pavilion which was commissioned in 2015 [21]. Their microgrid uses renewable energy, natural gas, and battery storage. It features three types of rooftop solar photovoltaic (PV) panel arrays—monocrystalline, polycrystalline, and micro AC inverter-based polycrystalline—that provide 50 kilowatts (kW) of electricity [21]. It also includes a 100 kW lithium-ion battery energy storage system (BESS) and a geothermal field with 15 wells drilled 500 feet (152 m) deep that serve heat pumps for heating and cooling systems [21]. Monitoring is achieved via a cloud-based analytics platform that collects system data and monitors each component of the system in real time [21]. Most operations of the microgrid can be accessed and monitored by using a smart phone.

Integrating renewable energy with diverse technologies creates a larger system that provides a range of benefits, including improved resiliency and generation flexibility [22]. Co-location of generation systems near loads reduces transmission and distribution costs leading to more sustainable electrical generation infrastructure. Microgrid generation can be deployed by using MW-scale individual technologies that are compatible and complementary using linkages to the existing distribution systems (see Figure 13.2).

Managing Multiple Renewable Microgrids

Smart microgrids require management. Management of multiple microgrids from a central location has the potential to reduce some operating costs. This can be accomplished with central microgrid digital control systems, advanced algorithms, artificial intelligence, and minimal human intervention. Predictive intelligence can be incorporated into the software and provide anticipated, self-adjusting responses to recurring events. Human intervention would be required only when unprogrammed responses to anomalies were to occur.

In such a scenario, more sophisticated communication and remote monitoring systems are required. The cost for these capabilities and microgrid management can be lessened when multiple smart microgrids are remotely monitored and controlled from a single central location. With multiple microgrids, each will likely have a different set of field objectives and different approaches to optimization. Maximizing interoperability requires linked communications among the microgrids when generating electricity [23]. The configuration hopefully reduces costs by reducing the time required to have technicians in the field. The system must demonstrate real-time or near-real-time control capabilities that incorporate optimization [23].

An example is the Model City Mannheim (Germany) project which included 1,000 households and multiple DERs. This study simulated a multiple renewable microgrid configuration from 2008 to 2013 and assessed the possibility of linking energy producers and consumers more closely using state-of-the-art information and communications technologies [24]. The goal was to connect electricity consumers either manually or automatically during low-price periods based on hour-to-hour variances in electricity prices to enable end-user savings [24]. The study proved consumer purchasing decisions changed in response to signals. Consumers shifted their electricity consumption to periods in which large quantities of energy from

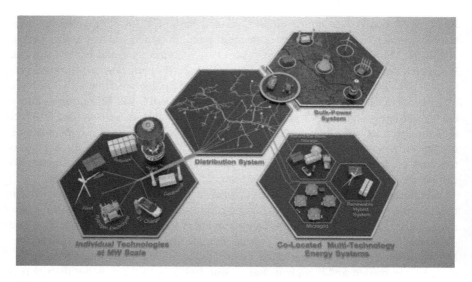

FIGURE 13.2 Integrating renewables in smart grids (J. Bauer, NREL [22]).

renewable sources were fed into the net and in which electricity supply volumes were high [24]. Overall electricity consumption remained low and electricity was relatively inexpensive during certain periods of the day. Consumers took advantage of the low-cost periods and programmed controllable household appliances to operate automatically [24].

POLICY CONCERNS

The development of smart microgrids in the U.S. faces multiple unresolved hurdles. Many have been identified for years but remain unresolved. This is primarily due to the lack of statutory regulation by state governments. In many parts of the U.S., it is the lobbying by incumbent industries and lack of enabling regulation that are the greatest factors hindering microgrid development.

Problems identified through microgrid case studies and interviews include establishing best practices for engaging customers, cybersecurity methods and protocols, interoperability standards, and microgrid communications and control systems [4]. Sandia National Laboratories has identified key developmental concerns including best practices for policies to: 1) implement dynamic pricing; 2) refine interconnection policies; 3) adjust retail rate designs and refine rates for partial-requirements service; 4) establish utility DER investment policies; 5) develop retail market participation rules; 6) provide utilities with appropriate regulatory incentives; 7) coordinate microgrid policies with other policies intended to promote DG, electric vehicles, and other distributed resources; and 8) achieve consistent regulatory policies across multiple utility service territories, including multiple-state, regional, and conceivably national policies [4].

Many smart grid development issues can be overcome with the implementation of improved policies supported by programs to incentivize microgrid research and local demonstration projects. Enabling legislation needs to be locally oriented and customized.

CONCLUSIONS

To satisfy customer needs and minimize investment in infrastructure, microgrids will need to become smarter and more reliable, and find ways to manage intermittent generation [3]. Smart microgrids are a subset of microgrids that use more sophisticated and robust technologies. Unlike central electric grids that must provide a pool of accessible power that is continuously available, smart microgrids offer an alternative as they use technologies that can better manage loads. The ability of smart microgrids to deliver electricity to loads on a targeted, as-needed basis is important to their success. The microgrid paradigm heavily relies on successfully providing this capability [3].

Smart microgrids are a form of advanced microgrids as they balance supply with demand in real time, dispatching resources that are capable of improving reliability. They accomplish this by optimizing energy profiles and efficiencies, enhancing island performance and maximizing the economic benefits associated with grid interconnectivity [4]. Advanced remote microgrids often operate in very isolated locations and often in harsh environments. The technologies they use must be hardened to mitigate environmental threats.

There are a number of key technologies that facilitate the development of smart microgrids. Sensor systems acquire information from the data environment and compile data that is of diagnostic interest. Electrical usage data from advanced metering infrastructure can be communicated directly to microgrid central controls. Intelligent electronic devices receive data from the sensors and power equipment and issue actionable control commands.

Smart microgrids employ advanced technologies and software to improve electrical system reliability and resilience. Another benefit of smart grids is their ability to lower the cost of electricity and services. They provide services that are improvements over traditional grid, offer more local control over services, increase local employment, and future-proof electrical supply systems.

Smart microgrids will also access and incorporate information about individual buildings, accessing their internal control systems, analyzing electrical energy data, matching supply availability with internal loads, and taking appropriate actions to reduce energy (and perhaps water) consumption within programmed limitations. They can be customized based on the complexity of the microgrid deployed and the internal capabilities of the building control systems. Internal access to building energy management systems enables microgrid operations to perform multi-tier custom diagnostics, and make selective decisions to schedule load shedding and level demand in real time. They are capable of using algorithms at multiple levels to provide engineering analysis and system updates. Reporting capabilities supporting energy managers will be available to the point of suggesting improvements to energy-consuming systems and modifications to buildings and other loads to optimize operations and system efficiencies (see Figure 13.3).

Deploying smart microgrids can be challenging. Some can be overcome with expanded use of digital control technologies. Policy challenges present another set

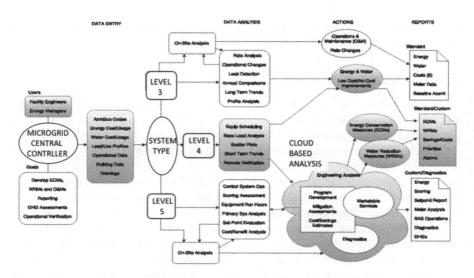

FIGURE 13.3 Schematic of microgrid software and control integration with energy management systems.

of issues. For example, in the U.S. there is a lack of statutory regulation by state governments. The need for enabling programs and resistance from incumbent industries are concerns. Regardless, smart microgrids are becoming more widely accepted as solutions to providing electrical services. There are many real-world examples of smart microgrid projects either under development or already deployed. Many are characterized by the use of mathematical modeling for components of microgrid DERs and the use of advanced software.

REFERENCES

1. Misra, G., Venkataramani, G., Gowrishankar, S., Ayyasam, E., and Ramalingam, V. (2017). Renewable energy based smart microgrids – a pathway to green port development. *Strategic Planning for Energy and the Environment*, 37(2), pages 17–32.
2. Gellings, C. (2015). *Smart Grid Planning and Implementation*. Fairmont Press, Inc.: Lilburn, Georgia.
3. Holdmann, G. and Asmus, P. (2019, September). What is a microgrid today? *Distributed Energy*, page 15.
4. Sandia National Laboratories (2014, March). The advanced microgrid, integration and interoperability. Sanida Report SAND2014-1535. https://www.energy.gov/sites/prod/files/2014/12/f19/AdvancedMicrogrid_Integration-Interoperability_March2014.pdf, accessed 24 November 2019.
5. Dwivedi, R. (2016, August 23). Smart grid-innovation ahead. https://ieeeuiettechscript.weebly.com, accessed 29 December 2019.
6. Strunz, K., Abbasi, E., and Huu, D. (2014). DC microgrid for wind and solar power integration. *IEEE Journal of Power Electronics*, 2, pages 115–126.
7. Shara, V. (2013, December 12). Research Gate. https://www.researchgate.net/post/What_is_the_difference_between_a_microgrid_and_a_smartgrid, accessed 4 November 2019.

8. Niclas (2019, February 25). Smart microgrids. https://sinovoltaics.com/learning-center/system-design/smart-microgrids, accessed 5 November 2019.
9. Mitra, J., Cai, N., Chow, M., Kamalasadan, S., Liu, W., Qiao, W., Singh, S., Srivastava, A., Srivastava, S., Venayagamoorthy, G., and Zhang, Z. Intelligent methods for smart microgrids. https://www.egr.msu.edu/~mitraj/research/pubs/proc/mitra-etal_intelligent_smart_mg_gm11.pdf, accessed 23 May 2020.
10. U.S. Department of Energy (2014, August). 2014 smart grid system report. https://www.smartgrid.gov/files/2014-Smart-Grid-System-Report.pdf, accessed 28 November 2019.
11. U.S. Department of Energy (2012, December). Application of automated controls for voltage and reactive power management – initial results. Smart grid investment grant program.
12. Galvin Electricity Initiative. Microgrid Hub. What types of technologies are used within a smart microgrid? http://www.galvinpower.org/resources/microgrid-hub/smart-microgrids-faq/types-technologies, accessed 23 May 2020.
13. Reilly, J., Hefner, A., Marchionini, B., and Joos, G. (2017, October 24). Microgrid controller standardization – approach, benefits and implementation. *Conference Presentation at Grid of the Future.* Cleveland, Ohio.
14. Grimley, M. and Farrell, J. (2016, March). ILSR's energy democracy initiative. https://ilsr.org/wp-content/uploads/downloads/2016/03/Report-Mighty-Microgrids-PDF-3_3_16.pdf, accessed 24 January 2020.
15. Jayant, K. and Alstom (2015, July). Microgrid system for Philadelphia Navy Yard, NREL – Advanced grid technology workshop series. http://www.nrel.gov/esi/pdfs/agct_day3_kumar.pdf.
16. Gerry and Sonia (2013). Optimal rural microgrid energy management using HOMER. *International Journal of Innovations in Engineering and Technology*, 2, pages 113–118.
17. Mariam, L., Basu, M., and Conlon, M. (2013). Community microgrid based on micro-wind generation system. *IEE Conference on Power and Engineering* (UPEC).
18. Buevich, M., Schnitzer, D., Escalada, T., Jacquiau-Chamski, A., and Rowe, A. (2014, July 8). Fine-grained remote monitoring, control and pre-paid electrical service in rural microgrids. *IEE Explore.* https://ieeexplore.ieee.org/document/6846736/authors#authors, accessed 20 November 2019.
19. International Electrotechnical Commission (2011). Electrical energy storage. http://www.iec.ch/whitepaper/pdf/iecWP-energystorage-LRen.pdf, page 74, accessed 17 January 2020.
20. Clancy, H. (2014, June 11). Honeywell joins forces with Fort Bragg on networked microgrid. https://www.greenbiz.com/blog/2014/06/11/fort-bragg-honeywell-join-forces-networked-microgrid, accessed 26 November 2019.
21. Black and Veatch (2019). Microgrid with energy storage system promotes reliability and sustainability. https://www.bv.com/projects/microgrid-energy-storage-system-promotes-reliability-and-sustainability, accessed 24 November 2019.
22. National Renewable Energy Laboratory (2019, November). Microgrids, infrastructure resilience, and advanced controls launchpad. DOE/GO-102019-5221. https://www.energy.gov/sites/prod/files/2019/11/f68/miraclefactsheet2019.pdf, accessed 20 December 2019.
23. Kroposki, B., et al. (2012). Summary report: DOE microgrid workshop. Report-out presentation, session 6: operational optimization. http://e2rg.com/microgrid-2012/Session-6_Report.pdf, accessed 27 November 2019.
24. MVV. The Model City Mannheim beacon project. https://www.mvv.de/en/mvv_energie_gruppe/nachhaltigkeit_2/nachhaltig_wirtschaften_1/innovationen_1/modellstadt_mannheim_1/moma.jsp, accessed 27 November 2019.

14 Scoping the Business Case for Microgrids

INTRODUCTION

There are a number of approaches to developing business models for microgrid projects. Many are similar to the ways that energy efficiency and alternative energy projects have been developed in the past. Drivers for microgrid projects include provision of electricity, reducing electricity and service costs, providing local development, accessing the benefits of renewables, and mitigating greenhouse gases. To develop a business model, there is a need to identify and quantify the value of the services microgrids provide. Goals of the business model include defining a project that is able to provide a positive cash flow and one that can be financed. This is confounded by the unique characteristics of each individual microgrid.

Despite the valuable outcomes that microgrids offer, there are difficulties in financing their development. The problem is not a lack of funding availability but often a lack of commercially available financing that is fenced for microgrid deployment [1]. Traditional financing approaches often fail to properly credit the annuity value of avoided future costs when future expenses, such as the costs of energy, can be widely volatile. Carbon taxation is used in some countries to mitigate such volatility but provides mixed results.

Additional hurdles include the difficulties in obtaining access to funds from local financial institutions. This can be caused by disconnects in lending practices as they apply to energy efficiency and carbon reduction improvement projects [1]. There may be structured restrictions on capital, collateral requirements, ownership provisions, and market conditions. This became increasingly evident in 2009, when lending was disrupted by a severe recession which impacted financial markets worldwide. There are ways to overcome such hurdles by using creative financial approaches. These include programs such as the use of carbon offsets and credits, funding demonstration projects, the use of performance contracting, and creative rule structures for measurement and verification (M&V) of savings. From 2009 to the beginning of 2020 the expanding world economy had ample investment capital available. However, economic conditions turned strongly recessionary in early 2020 due to the coronavirus (COVID-19) pandemic.

BUSINESS CASE FOR RENEWABLE ENERGY-BASED MICROGRIDS

Microgrid business models include single-user, multi-user, distribution company, or utility company ownership. Microgrids can operate using a single technology, multiple technologies, or hybrid and integrated technologies. Some include power storage systems and others do not. For microgrid development, project costs include capital costs, costs of maintenance, cost of operations, and other fixed and variable costs. The business case for microgrids must as a minimum address microgrid design, costs, project financing, creating a value proposition, and defining sustainable revenue streams. For microgrids with renewable generation, this means quantifying the value of the electricity (kWh), the value of demand (KW) projected, plus the values of any other ancillary other services to be delivered. These values vary over time and from site to site, confounding the process of lifecycle costing.

Costs and pricing structures fluctuate creating business risk. The cost of delivering renewable energy (RE) is continuing to decline. As costs decline, their use in microgrid configurations is likely to increase. Recent studies indicate that the costs of building and operating renewable-based energy systems have dropped below the operating costs of utility-scale coal and oil-fired electrical generation [2]. While this is promising, the deployment of renewables on their own is often unable to meet the needs of most developed countries. One of the best business case scenarios for microgrids is to use diversified and complementary conventional and alternative energy resources to meet the energy needs for the foreseeable future [2]. This is actually the trend that is occurring in practice. It is related to public policy approaches that support incremental improvements over a period of time and applies to community and military microgrid development.

The economics of each microgrid deployment is unique and has specific challenges and functional objectives [3]. The business case is typically justified using the economic value of the key project drivers. According to GTM Research, economic drivers that determine the market penetration of renewables include: 1) installation costs for renewables and energy storage; 2) fuel and power costs (and security of supply); 3) costs associated with outages; 4) deferral opportunities for transmission and distribution; and 5) project lifetime and cost of capital [3]. The costs of deploying systems to generated electricity are sensitive to capital costs, fuel costs, and subsidies. All must be considered in a lifecycle cost analysis to access the levelized cost of the energy generated. Using data from available research is one way to begin developing the business case for microgrids with minimal effort (see Figure 14.1).

Subsidies may be available to support project development. Energy resources and delivery systems are the most government-subsidized markets in the world. The greatest portion of these subsidies are directed toward fossil fuel development. These include discounted exploration rights, low-cost access to land, subsidized leases, direct funding, cost over-run support, environmental subsidies, government paid insurance support, and substantial tariffs on competing fuels. Subsidies are variable over time. They often increase the risks associated with developing lifecycle cost estimates as potential investors are uncertain as to how long subsidies will be available. For example, contracts for greenhouse gas (GHG) emission avoidance may have a term of only two years, making it difficult to help support projects that are financed over ten-year terms or longer.

When fossil fuels are used, carbon capture and sequestration (CCS) is worthy of consideration. It is a process that remains at the demonstration stage of the innovation process because of its high costs. However, international agencies (e.g., the IEA) are calling for further funding to support research and development or pilot project initiatives. The great expense of CCS implies that countries with many large point

Scoping the Business Case for Microgrids

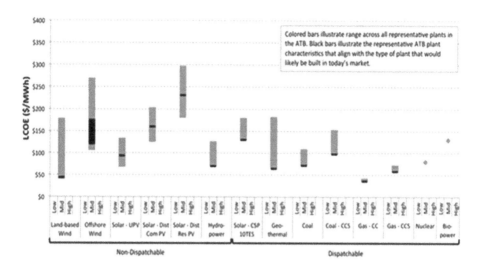

FIGURE 14.1 2017 levelized cost of energy for various electricity generation technologies (Source: NREL [44]).

source emitters (e.g., power plants and other industries) in close proximity to verified carbon storage sites are the most likely candidates for cost-effective CCS programs.

Key findings in a recent study by Lazard determined that alternative energy sources are becoming more competitive and are now at or below the marginal cost of certain conventional generation technologies [2]. Their study found that

"Global LCOE values for alternative energy technologies continue to decline, reflecting, among other things: 1) downward pressure on financing costs as a result of continuously evolving, and growing pools of capital being allocated to alternative energy; 2) declining capital expenditures per project resulting from decreased equipment costs; 3) increased competition among industry participants as markets evolve policies towards auctions and tenders for the procurement of alternative energy capacity (and away from standard offer programs, feed-in-tariffs, etc.); and 4) improving competencies in asset management and operation and maintenance execution [2].

Solar PV and onshore wind power are approaching the levelized cost of electricity when compared to conventional coal-fired combined cycle plants (see Table 14.1). Geothermal plants are capable of generating baseload electricity and already favorably compete with coal.

CREATIVE APPROACHES TO PROJECT FINANCING

Much of the early financing for microgrid development in the U.S. was provided by government agencies. Today, a private financing market is emerging through large companies, local banks, and utilities. Microgrid financing is often provided for all project components, such as power and energy generation systems, and any associated energy efficiency retrofits, distribution systems, central control systems, and

TABLE 14.1
Unsubsidized Levelized Cost of Electricity (LCOE) for Resources Entering Service in 2020 Using 2017$s/MWh (Source: EIA [5])

Plant type	Capacity factor (%)	Levelized capital cost	Levelized fixed O&M cost	Levelized variable O&M cost	Levelized transmission cost	Total system costs LCOE
Renewables						
Wind (onshore)	41	43.1	13.4	0.0	2.5	59.0
Wind (offshore)	45	115.8	19.9	0.0	2.3	138.0
Solar PV	29	51.2	8.7	0.0	3.3	63.2
Solar thermal	25	128.4	32.6	0.0	4.1	165.1
Hydroelectric	64	48.2	9.8	1.8	1.9	61.7
Biomass	86	39.2	15.4	39.6	1.1	95.3
Geothermal	90	30.1	13.2	0.0	1.3	44.6
Fossil fuels with present technologies						
Coal using 30% CCS	85	84.0	9.4	35.6	1.1	130.1
Conventional combined cycle	87	12.6	1.5	34.9	1.1	50.1
Advanced combined cycle	87	14.4	1.3	32.2	1.1	49.0
Advanced combined cycle with CCS	87	26.9	4.4	42.5	1.1	74.9
Conventional combustion turbine	30	37.2	6.7	51.6	3.2	98.7
Advanced combustion turbine	30	23.6	2.6	55.7	3.2	85.1
Advanced nuclear	90	69.3	12.9	9.3	1.1	92.6

Source: USEIA, International Energy Statistics, as of August 8, 2018.

connections [6]. Developers of microgrid projects can improve project bankability if they include both electricity and thermal energy as part of a project [6].

Hybrid systems complicate the financing of microgrid projects. This is due to their perceived novelty, the lack of comparative development, difficulties in predicting the costs for development, and variable pricing for their products. Regardless, reductions in costs and improvements in capabilities of renewable energy generation systems, energy storage, along with advances in technology and communications, enable economic justification of hybrid applications that previously would have required special support or incentives [7]. The integration of RE technologies can reduce operating expenses when compared to fossil fuel-fired generation, while optimizing system reliability, efficiency, and flexibility [7]. Since 2000, diesel fuel prices have mostly trended upward while wind power prices trended slightly downward and solar PV prices have fallen dramatically (projections place the average price of wind energy at US$0.09/kWh in 2020, and the price of solar energy only slightly higher) [7]. In other regions of the world, solar PV can be supplied at less than $.03/kWh. Downward trends in prices and reduced volatility give RE a long-term price advantage over conventional diesel-generated electricity.

CAP AND TRADE PROGRAMS

The report "The Economics of Climate Change" explains how cap-and-trade systems control the overall quantity of GHG and CO_2 emissions by establishing binding emissions commitments [8]. These systems regulate contaminates from generation sources by establishing limits on the allowable levels of emissions and by requiring reductions in contaminate levels over a period of time.

Using cap-and-trade systems, an emission maximum is established for the pool of participants and each participant in the pool. Within this specified emission quantity ceiling, entities covered by the plan (e.g., individuals, organizations, industries, or countries) are then free to choose how best, and where, to deliver emission reductions (Figure 14.2). The largest example of a regulated cap-and-trade scheme for GHG and CO_2 emissions is the EU's Emissions Trading Scheme.

There are a number of examples of regulated and voluntary national and regional emissions trading schemes, including the former Chicago Climate Exchange (CCX) and the U.S. Regional Greenhouse Gas Initiative (RGGI). The Chicago Climate Exchange provided an example of a voluntary exchange. Now inactive, it allowed participants (e.g., companies who had accepted voluntary commitments to reduce emissions) to purchase carbon financial instruments (CFIs). Projects eligible for CFIs included reforestation, afforestation, methane mitigation, renewable energy generation, and offsets in the agricultural and biomass sectors. Offsets were created through the use of conservation tillage and grass planting. Eligible projects were enrolled through an intermediary registered with the CCX, which served as an administrative and trading representative on behalf of the multiple individual participants. The intermediary body was called an *offset aggregator*. The first sale of an exchange of verified CO_2 offsets generated from agricultural soil sequestration in the U.S. occurred in April 2005 [9].

By June 2006, approximately 140,640 hectares (350,000 acres) of conservation tillage and grass plantings had been enrolled in the U.S. states of Kansas, Nebraska, Iowa, and Missouri [11]. The measures to enhance natural soil fertility and carbon sequestration potential also decrease the use of human-made fertilizers, reduce deforestation, improve water quality, reduce power consumption, and lessen fuel requirements for tillage [12]. While it seems counterintuitive, green space on land used for microgrids might be eligible in addition to offsets from renewables.

An example of a regulated regional U.S. cap-and-trade system that applies carbon offsets is the Regional Greenhouse Gas Initiative, an initiative developed by states in the Northeast and Mid-Atlantic regions. The basic objective of the RGGI is to reduce GHG emissions from the electrical power production sector while maintaining economic competitiveness and efficiency across the regional power market. The framework of the RGGI follows the model of earlier successful cap-and-trade systems, such as those used to reduce acid rain-producing sulfur dioxide (SO_2) emissions. Every regulated cap-and-trade program includes the following basic components:

- a decision about who is being regulated
- mandated emissions reduction levels
- provision for the distribution or allocation of permits or allowances that power generators require
- a structure for the trading mechanisms

The RGGI agreement regulates medium and large electrical power plants. Nearly 75% of the allowances in each participating state are distributed among the generators, while the balance is set aside for various public benefits. The agreement also allows the generators who reduce CO_2 and other GHG emissions further than required to bank those allowances for the future or sell them to other generators. Moreover, generators who have not met their reduction goals must purchase allowances on the

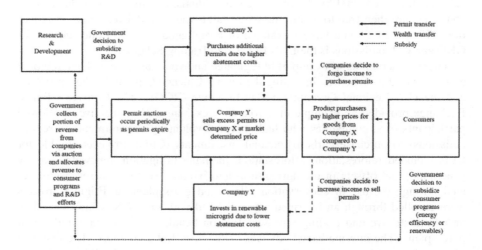

FIGURE 14.2 How cap-and-trade programs work (Source: adapted from [10]).

market or obtain a limited amount of reductions through offsets. Carbon offsets are CFIs that represent emission reductions that are typically achieved outside of the power production sector. These offsets are financial tools that provide incentives for emitters of CO_2 to reduce their emissions. Offsets are typically derived from emission reductions through mitigation efforts such as energy conservation, alternative energy projects, reforestation, or landfill methane recapture. The RGGI capped CO_2 emissions from electrical power plants at 2005 levels between 2009 and 2014. After 2015, the cap decreased, so that by 2019, the states collectively reduced their emissions to at least 10% below 2005 levels [13].

RENEWABLE ENERGY CERTIFICATES

Green certificates, also called green tags, renewable energy credits (RECs), green certificates, or tradable renewable certificates (TRCs), are tradable, non-tangible energy commodities that in the U.S. represent proof that one megawatt hour (MWh) of electricity was generated from an eligible RE resource and was fed into the shared system of power lines which transport and distribute electricity [14]. The REC certifies that the RE generator has supplied the electricity and it has been fed into the host grid. Most are sold as independent devices separate from the actual electricity delivered. These certificates can be sold, traded, or bartered, and the owner of the REC can claim to have purchased the renewably sourced electricity [14]. RECs may be purchased by utility companies to meet the requirements of a renewable portfolio standard (RPS) or voluntarily purchased by individuals of companies. The certification process usually requires third-party verification by an independent auditor. Green tags are also used as offset mechanisms to mitigate the amount of CO_2 generated on one's behalf by purchasing them to offset carbon emissions [15].

GLOBAL ENVIRONMENTAL FACILITY AND CLEAN DEVELOPMENT MECHANISM

To ultimately reduce the global cost of stabilizing GHG in the atmosphere, investments in fast-growing economies must incorporate energy efficiency and low-carbon technology [11]. These technologies include those that microgrids deploy to produce renewable electricity. With the exception of military microgrid development, private sector resources for microgrids far outweigh the total funds available from governments and multi-lateral institutions. Middle-income countries, where the bulk of future GHG emissions growth is concentrated, generally have access to capital from the private sector. Public sector resources are also an important lever to channel flows of domestic and international private sector investment into carbon reduction technologies.

Successful and effective financial solutions to supply renewable energy and reduce carbon emissions include programs advanced by the Global Environmental Facility (GEF) and the Clean Development Mechanism (CDM). The GEF was established to help tackle pressing environmental problems. It has provided close to $20 billion in grants and mobilized an additional $107 billion in co-financing for more than 4,700 projects in 170 countries [16]. The GEF has a history of financing energy efficiency and renewable energy projects especially in developing countries. Many of the most profitable projects to generate carbon credits under Kyoto's CDM, such as methane gas projects, were exhausted after a flurry by Western financial houses to secure them. The CDM provides an important channel for private sector participation in financing low-carbon investments in developing nations. The two primary requirements of CDM projects is that they contribute both to the reduction of emissions according to a baseline or predetermined scenario, and to sustainable development according to priorities and strategies defined by the host country [17]. While many lucrative projects reflect activity to reduce emissions of the most potent greenhouse gases, there is concern that some long-term solutions needed to fight climate change are not being financed. The collective impact of these programs is small relative to the challenge of reducing CO_2 and GHG emissions.

The World Bank has recently suggested that the GEF could play an enhanced role in encouraging technology transfer and lowering the cost of the low-carbon technologies that are relevant to the priorities of developing countries. To continue to expand over the next decade, the GEF would require at least a two-fold increase in current financing to ensure sustained market penetration of energy efficiency and RE technologies. Financing a strategic global program to support the reduction in costs of pre-commercial, lower GHG emitting technologies such as carbon capture and sequestration (CCS), solar thermal, or fuel cells would be much greater, requiring more than a ten-fold increase.

Carbon offset credits are tradable credits earned for investing in projects to reduce greenhouse gas emissions. In CDM terms, these credits are called certified emissions reductions (CERs) and are generated in developing countries. One CER represents an emission reduction on one metric ton of CO_2. Such credits are valuable to governments and companies in developed nations because the Kyoto Protocol allows them to use these credits to offset their own GHG emissions and help them meet a portion of their domestic reduction targets [18]. Buying credits, or paying for them to be generated, can in some cases be less expensive than reducing emissions at home. In 2005, 374 million tCO2e, mainly of CERs, were transacted at a value of US$2.7 billion, reflecting an increase of more than three times the previous year's volumes from project-based transactions [19]. While over $1.5 billion annually were generated by the end of 2012, the prices for CERs dropped from $20 per metric ton in August 2008 to $5 per metric ton by September 2012, when the targets under Kyoto expired [20].

The CDM program has been criticized for rules which create opportunities for investments in easy projects at the expense of more expensive projects to implement renewable energy. Many want more RE projects to be implemented in developing countries to encourage sustainable economic development in less wealthy regions of the world. For example, one of the largest CDM transactions (300 million euros) was for a hydrofluorocarbon (HFC) destruction project in China at Zhejiang Juhua chemical plant generating 4.8 tCO_2-equivalent reductions annually [21]. After opportunities for HFC destruction were mostly exhausted, many newer CDM projects have since focused on mitigation and switching from fossil fuels to RE.

BUILD-TRANSFER AGREEMENTS

Build-transfer agreements are those in which a utility contracts with a third-party project developer to design, oversee development, and construct a project. These

Scoping the Business Case for Microgrids 225

can be used to develop distributed electrical generation infrastructure. When the project is nearing completion or perhaps complete, ownership of the project is transferred to the utility. A build-transfer agreement is a hybrid form of procurement contract which has components similar to both acquisition agreements and construction contracts [22]. The developer's role is to secure the needed land rights, permits, interconnection rights, and project contracts [22]. The developer then develops the project for the utility with ownership transferred to the utility before the project is commissioned and begins commercial operation [22]. In the U.S., the host utility often accepts ownership of the project before it is placed in service to reduce federal tax exposure. After initial startup, the project may be operated and maintained by the utility or a third party [22].

Peer-to-Peer Trading

Implementing new business models for developers of renewable energy projects has potential to disrupt the markets for delivering electricity. There are opportunities which would make microgrids the preferred way to deliver power in the future if interconnections are secure and load-balancing strategies are successful. Among those available for community-based microgrids is peer-to-peer (P2P) energy trading. This model allows for the direct transfer of electricity with immediate payment by the consumer. To obtain the best electrical rates, customers might forward their meter data to several potential electrical suppliers with a goal of reducing costs by obtaining bids for the energy used. By compiling numerous successful bids, energy suppliers could create bankable projects. However, there are misconceptions as to how P2P trading can be used to support developers in their efforts to obtain financing for microgrid development. It is often interpreted as the marketing of environmental attributes such as green tags, green certificates, or RECs which is inaccurate [23]. For microgrids, it is actually a direct pathway for the transfer of electrical energy from a distributed energy source to the consumer plus a means to reduce the soft costs associated with microgrid development.

The business model for P2P often involves eliminating the transfer agents and their fees. Blockchain technologies can be used for this purpose. They can improve and expedite accounting for the transfers. Using a blockchain-based marker each distributed energy resource can be assigned a digital identity linking each owner of the corresponding credits [23]. Each customer would have a digital signature enabling them to order energy upon demand and have digital metering systems. The blockchain would manage the accounting for these transfers. The blockchain networks could also be used deliver new services and products. These would include monetizing data streams, better integrating and managing demand-side resources, and helping to launch services to finance and own energy infrastructure [23].

WAYS TO COMBINE PROJECT FINANCING AND DELIVERY

Governments in the U.S., particularly local and state ones, have been tasked with finding ways to implement sustainable development policies and programs. With scarce funding, inadequate resources, and limited technical skills, this has often proved challenging. Examples of ways to overcome such inertia include using public–private partnerships and energy savings performance contracts. Both of these tools have been used successfully to develop microgrids.

Public–Private Partnerships

Sustainability agendas can be successfully achieved by using public–private partnerships (PPPs). In the past, public–private partnerships have often focused on issues such as environmental concerns, local development, patterns of energy use, water, tourism, and education [24]. All can be linked to local sustainability. With a rich history of using PPPs, many governments are skilled at using this model successfully. Today, community microgrids are being developed using PPPs. Such partnerships enhance local microgrid development of fostering cooperation among the stakeholders. The San Diego Gas and Electric (Borrego Springs) microgrid in California is one such example.

A substantial body of literature is available concerning public–private partnerships. Most focus on diverse development projects. The literature considers sources of risk capital for governments, public incentives, the relationships between governments and the private sector, the types of partnerships, the results and assessments of PPPs, and ways they can be improved [24].

There are four primary sources of risk capital: 1) the private, for-profit sector; 2) governments; 3) employee savings and benefit funds; and 4) the private, non-profit sector [25]. Each source of funds has its benefits and costs. For-profit enterprises consider the advantages based on management efficiency, acceptable rates of return on capital, net margins, and the ability to provide adequate incentives to skilled workers as critical to their success. Alternatively, governments tend to "avoid risks, invest insufficiently and avoid cost reduction measures" [25]. According to Clarke and Gayle, the idea of public entrepreneurial activities "seems an oxymoron: the risk-taking business actors who spot new opportunities for making profits and challenge the status quo" operate in an environment controlled by "risk-averse politicians seen as dominating local government" [26]. In public–private partnerships, the potential partners often seek much different sets of goals, which creates ample opportunities for friction and conflicts of interest.

To attract capital, public incentives and inducements are often used. Schneider suggests that, "as sources of competition, special districts, county agencies and private firms can all increase their efficiency and responsiveness in the provision of public services" [27]. He also concludes that "increasing local government efficiency allows more services to be delivered per tax dollar" expended; however, attempts "to increase local government efficiency have encountered difficulties, chiefly because we cannot monitor and improve something we cannot measure" [27]. Energy savings provides an example since calculating project savings over a period of time involves measuring the absence of expenditures. Such difficulties are an example of the issues that hamper the ability of governments to implement efficiency opportunities. How can efficiency opportunities be encouraged? One approach is to implement monitoring schemes with an appropriate M&V regimen.

A product or public good can be provided by either the public sector, the private sector, or by a partnership between both. The provision of electricity offers

Scoping the Business Case for Microgrids

an example. There are various types of public–private partnerships differentiated by the ways they are structured. The types of possible relationships between governmental entities and the private sector to provide services include competition, co-production, provision, for-profit government-owned activities (quasi-governmental organizations) using private subcontractors, independent government-owned/contractor-operated (GOCO), and government-regulated for-profit corporations [24]. There are also instances of more than one governmental entity (local and federal) being in partnership with multiple private sector players to deliver a project or public good.

Competitive postures include services or products that can be provided by either the public sector or the private sector when competing for the same market. This occurs in the provision of education or housing. Universities may be public, private, or a combination of the two. The U.S. Postal Service is a quasi-governmental, for-profit (theoretically) service competing directly with for-profit companies for some delivery services in overlapping markets. Convention centers are another example of facilities owned and operated as a public good. The Atlantic City Convention Center is New Jersey's largest conference center. The facility completed a microgrid project which when initially completed included the largest roof-mounted, grid-connected PV array in the U.S. (see Figure 14.3). Performed in partnership with Pepco Energy Services, the project generates about a quarter of the electricity required by the facility and saves an average of $220,000 annually over its project life.

There are numerous examples of public–private relationships. Regulated public utilities with their geographically designated and protected service territories are a form of government regulated, for-profit corporations [24]. Public entities such as municipal utilities might provide electrical power to one market while a private regulated utility provides power to another. Co-production of goods is exemplified by GOCO facilities, particularly military munitions, infrastructure, and hardware. In non-competitive arrangements, services may be co-produced by means of PPPs. Public–private research and development activities are another example of a more cooperative posture [24]. The non-competitive arena includes redevelopment. "The common interest of both government and developers in maximizing the potential returns on a site allows flourishing PPPs. Most of these arrangements involve joint public–private participation in major commercial projects" [25]. While commercial developers benefit from the relaxation of regulations, inducements to invest, and infrastructure enhancements, the governmental entities involved benefit from the investment in the community, increased local employment, an increased property tax base, and the prestige associated with the projects themselves. Local officials have become more entrepreneurial, partnering to build public parking garages and selling or leasing facilities to retailers [28].

Increased public–private cooperation is one result of these often complex partnerships. While there are many benefits to PPPs, difficulties do arise. Locally initiated PPP arrangements are non-standardized, diverse in nature, and may require enabling legislation and/or administrative regulations to effect. A driver from the local government perspective is that in-house technical skills are not always available in the public sector or are too costly to procure. This causes make-versus-buy decisions to be unlikely since the governmental entity lacks the needed project support

FIGURE 14.3 Rooftop solar array at the Atlantic City Convention Center.

capabilities. In such cases, there is cause to seek technical skills from private sector companies capable of supplying the expertise required. This creates opportunities for microgrid developers with a history of successful projects. Other drivers from a local government's perspective are that development projects create high-paying employment during their construction and a non-public partner may be able to expedite project completion. The sooner projects can be completed, the sooner economic benefits can become available.

An important concern of the public sector is financial default by a private sector partner. Corporate entities have a greater tendency to be short-lived compared to governments. What happens when the for-profit entity defaults on its obligations or files bankruptcy? How is the governmental entity contractually protected from such defaults? Contract clarity and third-party insurance are often the solution.

Despite difficulties and discordant agendas, there are those in both the governmental and private sectors who advocate strongly in favor of public–private partnerships. Key components leading to successful PPPs have been suggested. Blakely believes that they can create "bridges of trust based on similar objectives" and for those seeking successful partnerships, he offers the following formula [28]:

- a positive civic culture that encourages citizen participation and long-term employment in the community
- a realistic and accepted community vision that considers the community's strengths, weaknesses, and potential

- effective civic organizations that can blend member self-interest with broader community interests
- a network of groups and individuals that encourages communication and fosters mediation of differences among competing interests
- the means and desire to encourage civic entrepreneurship (risk-taking)
- continuity of policy, including the ability to adapt to changes and reduce uncertainty for businesses [28]

ENERGY SAVINGS PERFORMANCE CONTRACTS

Energy service companies (ESCOs) have been in existence for over a century. They are companies that are normally associated with identifying and implementing cost-effective energy efficiency upgrades to facilities. Energy savings performance contracts (ESPC) are a means of implementing and financing energy efficiency upgrades by partnering with an ESCO that guarantees savings that result from upgrading building systems. In an ESPC, the ESCO assumes some or all of the financial risk of the investment by guaranteeing that the host organization (e.g., a university, school district, governmental entity, industrial concern, hospital) will realize the anticipated savings over the contract term. This model has been successfully used for military and community microgrid development.

Energy savings performance contracts are actually a type of financial mechanism used to implement a project. These projects traditionally begin with an energy and/or water reduction assessment to consider the types of improvements that would reduce or eliminate recurring utility costs in existing facilities. Working on the customer side of the electric meter, the assessors provide a list of opportunities, identifying the scale and scope of the various improvements available to reduce operating costs. Facility owners use the preliminary assessment to estimate the energy savings and financial impacts of potential investments in efficiency and alternative energy improvements. Those improvements selected are packaged to provide economic justification for project implementation by the ESCO. In some cases, the ESCO will also assume the risk of financing the improvements themselves.

ESPCs rely on subsidizing projects with utility savings and other cost avoidance, using M&V procedures to document savings, and the provision of a third-party guarantee of savings if needed. ESPCs provide an important mechanism for many cash-strapped organizations to upgrade building systems and equipment. From the building owner's perspective, future recurring utility costs are reduced by the improved efficiencies of the newer or upgraded equipment. In the case of microgrids, the value of the electricity generated, electric demand costs avoided, and other services provide quantifiable benefits. The guaranteed reductions in future utility costs are used to finance the improvements. Any shortfalls in projected savings during the contract term are covered by the savings guarantee.

The energy services provider pays for the implementation of the project and manages the M&V process [29]. ESPC projects are structured follows:

1. An ESCO is selected.
2. An agreement is signed to proceed with the planning phase on the project.

3. An investment-grade audit is performed.
4. The owner approves the measures to be implemented.
5. An ESPC is entered into by the owner and the ESCO and financing is secured for the upgrades.
6. A turnkey project is implemented under the oversight of the ESCO.
7. Operations and maintenance procedures are established to manage the efficiency of the facilities.
8. Measurement and verification plans are developed with procedures in accordance with established engineering protocols. Baseline measurements obtained during the energy assessment phase are used for savings comparisons.
9. Payments are made to the ESCO and the financier.
10. Annual reconciliations are performed to determine actual savings. If savings are not realized, the ESCO is responsible for paying the owner the difference between the guaranteed savings and the actual documented savings [30].

Large investments in ESPC projects are occurring. The U.S. federal government has entered into over 400 performance contracts worth about $1.9 billion dollars (using private sector investment—$3.5 billion including financing), to guarantee energy savings of $5.2 billion through reductions in utility costs [31]. The net benefit to the government from these projects has totaled over $1.7 billion dollars [31]. Local governments, such as the cities of Baltimore, Maryland and Bloomington, Indiana, have successfully used this process. The City of Covington, Kentucky used an ESPC program to qualify a guaranteed savings of $2.25 million over project life [32]. Using the ESPC model, the U.S. Marine Corps Logistics Base in Albany, Georgia developed a 15.6 MW capacity microgrid project at a cost of $4.2 million in 2018 [33]. It includes 8.5 MW of biogas electric generation, a 4.1 MW landfill gas component, plus six diesel generators with a capacity of 3 MW [33].

The ESPC investment model can be used to implement carbon reduction improvements in facilities and processes. ESPC projects that improve the efficiencies of using carbon-based fuels or use substitute forms of energy (such as solar or wind energy projects) provide carbon mitigation potential. Software tools that analyze and assess the savings from these projects are capable of simultaneously calculating the carbon emission offsets from these projects. The documentation generated from the analysis can be used to calculate the value of carbon offsets available from cap-and-trade regimes in local currencies. The market value of the offsets generated also provides yet another revenue stream for financing microgrid projects. In the U.S., tax credits are available for certain types of carbon capture and sequestration projects.

CHALLENGES TO MICROGRID DEVELOPMENT

The business case for microgrids tends to improve as the costs of distributed generation and energy storage decline. Regardless, policy and political pressures inhibit the deployment of microgrids [34]. A fundamental issue is the frailty of the microgrid business model itself. Many microgrid development projects are custom one-off

Scoping the Business Case for Microgrids

opportunities that often defy scalability. There is also the fundamental development issue that most microgrids start small and grow incrementally. For example, an initial project scope to generate electricity using solar PV and a fuel cell might later add battery storage.

POLICIES IN FLUX

As the expansion of microgrid infrastructure occurs over time, the policies that apply to their development are evolving. This means that standards and regulations that apply at the beginning of a microgrid development may change when the microgrid is expanded at a future date. The economic benefits associated with a microgrid can also change over time. For example, the customers of a remote microgrid might at a future date be offered lower cost electricity by a central grid extension which causes microgrid assets to become stranded. This is a worst-case scenario that tempers investment in microgrids due to their associated business risk.

DEVELOPMENT AND SCALABILITY ISSUES

The need for custom project design and delivery creates problems for the development and financing of microgrids. Robyn Beavers, vice president at Lennar Technology and Investments, once stated that there is no "truly scalable platform where a microgrid is a cost-effective energy infrastructure alternative… there is a technology and it works, but there is not a front-end business model or the back-end financing" [35]. Partnerships with the host utility can overcome this problem. Utility ownership rates of microgrids approached 18% in 2016. However, regulatory approvals are challenging and there is debate as to whether or not the costs should be absorbed by the utility's entire customer base or just the microgrid's interconnected customers [35].

Larger microgrids such as those that connect multiple centers and cross public rights-of-way face multiple policy barriers that often require time to resolve and add to development costs. These have been categorized by Gimley and Farrel [34] as:

- Microgrids are undefined in most state laws, leaving legal uncertainty about whether they will be classified and regulated as utilities.
- Microgrids may run afoul of state laws prohibiting the sale of electricity by non-utility entities.
- Microgrids may have particularly challenging interconnection issues that are not addressed by single generator interconnection rules.
- Microgrid central controller systems have typically been individualized, inhibiting easy replication and increasing cost.
- Microgrids may have few revenue streams if there aren't markets for services such as voltage control and frequency regulation.
- Due to the various microgrid ownership models (the utility, a public entity, a private company, or a partnership combining these entities), it is sometimes unclear how benefits should be shared and costs allocated.

Controller Technologies

The business case for a microgrid is bracketed by the functional capabilities of the controller. Many consider the concept of the microgrid as being enabled by the controller and defined as the resources—generation, storage, and loads—within boundaries that are directly managed by the controller and its capabilities [36]. The controller "manages the resources within the microgrid's boundaries, at the point of interconnection with the utility, interacting with the utility during normal operations" while defining "the microgrid's operational relationship with the distribution utility" [36].

It can be a problem finding controllers to fully utilize the customer-based resources [34]. A solution is to construct a project-specific microgrid visualizer to enable the utility to monitor all operations [34]. This can be costly to implement. When successful, microgrid systems integrate software and control systems, such as smart meters, to manage the grid operation in an efficient and reliable manner [37].

Electricity Pricing

Utility pricing and billing structures may vary based on a set of defined parameters and circumstances under which microgrid resources are called upon and could reflect incentive-based performance mechanisms based on the quality of the services provided [38]. These include variable demand-response pricing models, responsibility for operations and maintenance, peak demand charge avoidance, and fluctuations in fuel costs that are a function of microgrid system configuration [38].

There are innovative approaches to pricing electricity generated by microgrids that optimize energy scheduling with renewable energy systems by using smart tariffing approaches [39]. Integrating RE generation within microgrids becomes a more viable option in locations with increasing electricity rates or volatile fuel prices [3]. A novel approach to electrical supply and pricing is the clean peak standard (CPS) which is being considered in several U.S. states. Similar to a renewable portfolio standard (RPS) which typically identifies a percentage of electricity which must be provided to the grid by the host utility, the CPS provides a pricing premium for electricity generated by renewable energy [40]. This premium would apply during periods of peak electrical load when prices for electricity or greenhouse gas emissions are higher due to demand changes [40]. The pricing program qualifies the electricity supplied and defines when the rate structure applies. CPS programs can incentivize hybrid systems that use renewables, energy storage, or demand reduction strategies, and provide opportunities for additional revenue for microgrid owners [40]. Presently, there is no template for developing specific tariff programs and rate structures for microgrids that enable the distribution operator to recover costs and allow a retail supply of microgrid-based generation and storage services [41].

The pricing of externalities associated with greenhouse gases is rare in the utility industry. There are markets for carbon emissions. U.S.-regulated markets include California's AB32 and the Regional Greenhouse Initiative which serves ten New England states. In many markets, the lack of externality pricing allows the market to ignore the full costs of producing grid-supplied electricity [41]. This hidden subsidy is a disadvantage

Scoping the Business Case for Microgrids 233

for renewable energy-based microgrids. Microgrids can expand their building control systems and demand response programs, which improves the value proposition when externalities are considered [41]. To make matters more difficult, few policies currently in place subsidize these or other benefits of microgrid deployment. Given such limitations, the value proposition to deploy a microgrid typically includes [33]:

- renewable energy integration
- resiliency
- utility cost savings and demand charge abatement
- reliability
- power quality improvement
- reduction in carbon footprint
- provision of energy and capacity services
- provision of ancillary services
- linkage to virtual power plants
- future transactive energy revenue
- non-electricity services (thermal, water, etc.) [33]

COST OF GENERATION SYSTEMS, INTEGRATION, AND MAINTENANCE

Microgrids present opportunities for local small and medium size generation. Perhaps the greatest barrier to microgrid deployment has to do with the economics of scale associated with their functional deployment and fixed costs. The design and analysis of microgrids is often a custom activity which increases development costs. The goal of plug-and-play components and system architecture has not yet been achieved, meaning that there are likely additional design, engineering, and integration costs.

Microgrid development costs vary based on location and market segment. There are six different levels of microgrid sophistication, starting with the simplest which has only a backup generator to the most advanced microgrid with multiple types of generation, electricity or thermal energy storage, sophisticated controller capabilities, weather and generation forecasting, and the ability to coordinate multiple microgrids [42]. Identifying the type of configuration and application of a proposed microgrid helps to establish the potential development or expansion costs (see Table 14.2). For example, a Level 3 configuration for a 40 MW campus application might have system components that include a controller, CHP, solar PV, and energy storage. It is a common configuration for most microgrids except for those associated for military applications.

While the costs for microgrids vary, engineers find it helpful to consider the average costs of previous projects based on application, scope, and capacity. A study of 26 microgrid projects provided an average installed cost of roughly $3.1 million per MW [33]. For example, the Eagle Picher microgrid in Joplin, Missouri includes 20 kW of solar PV, 10 kW of wind power, plus a battery storage system. With a total capacity of 1.03 MW total costs were $2.6 million [33]. Costs for microgrid development tends to be higher in North America than elsewhere in the world. This is due to the higher costs of land, labor, and technical skills compared to other regions of the world. An analysis of total microgrid costs per MW commissioned by the National Renewable Energy Laboratory (NREL) shows that the community microgrid market

TABLE 14.2
Microgrid Levels of Configuration and Common Applications [43]

Types of microgrid configurations	Characteristics	Utility	Campus	Commercial	Industrial	Military
Potential microgrid						
Level 1	Stand-alone generator	Yes				
Level 2	Multiple distributed generation units	Yes	Yes			
Coordinated microgrid						
Level 3	Controller, thermal assets, renewables and energy storage	Yes	Yes	Yes	Yes	
Level 4	Level 3 assets + controller, load management	Yes	Yes	Yes	Yes	Yes
Level 5	Level 4 assets + controller, forcasting, economic dispatch	Yes	Yes	Yes	Yes	Yes
Level 6	Coordination and control of multiple microgrids	Yes				

has the lowest mean installed cost at $2.1 million/MW, followed by the utility and campus markets, which have mean costs of $2.6 million/MW and $3.3 million/MW, respectively (see Table 14.3) [43]. Finally, the commercial market has the highest average cost, over $4 million/MW [3].

Energy storage systems represent an added feature which augments project costs. They may require an independent cost/benefit analysis to be economically justifiable. The technological and commercial maturity of energy storage types and components vary widely [44]. Some systems, such as lead-acid batteries and sodium-sulfur batteries, are proven technologies with many years of maturity while others, such as flow batteries and emerging lithium-ion batteries, are newer and have a more limited operational maturity [44]. While capital cost is important, a lifecycle cost analysis and or a comparison of cost per delivered kWh over the project life is an equally important business case evaluation metric [44]. Table 14.4 provides information on storage system capacities, efficiencies, and costs. Pumped hydropower and CAES offer the lowest storage costs, largest capacities, and greatest longevity for microgrid applications.

The system integration process (including microgrid electrical, control, and security subsystems assembly, implementation, and operations) has been identified as a key integration-related challenge to the point of being cited as potentially cost-prohibitive [45]. In some cases, costs for system integration can be as much as one-third of the total project costs [45].

The cost of system operation and maintenance (O&M) must not be overlooked. Components of O&M costs can be categorized as either variable or fixed. For fossil fuel-fired electrical generation systems, the largest share of variable cost is due to the volatility of fuel costs. Since renewable energy systems do not usually require fuel (exceptions include biomass), they have much lower long-term variable costs. Comparative variable and fixed operations and maintenance costs are shown in Figure 14.4. O&M costs for most of the microgrid-related technologies are manageable. However, landfill gas generation projects have by far the highest O&M costs.

CASH FLOW ANALYSIS FOR A MICROGRID PROJECT

From a business perspective, the ultimate goal is to develop a robust case to support development of the microgrid and qualify the project for long-term financing. To be marketable, the proposal needs to assign value and quantify project benefits (cash inflows) and costs (cash outflows) including project financing. If a 5 MW utility

TABLE 14.3
Market Segment Microgrid Costs in Million $/MW [44]

Market segment	Interquartile range	Mean
Campus/institutional	$4.94 to $2.47	$3.34
Commercial/industrial	$5.35 to $3.40	$4.08
Community	$3.34 to $1.43	$2.12
Utility	$3.22 to $12.32	$2.55

TABLE 14.4
Energy Storage System Cost (Source: EPRI, Adapted From [45])

Application Storage technology	Capacity (MWh)	Power (MW)	Duration (hours)	% Efficiency (total cycles)	Total cost ($/kW)	Cost ($/kwh)
Bulk energy storage w/RE integation						
Pumped hydropower (microgrid scale)	1,680–5,300	280–530	6–10	80–82 (>13,000)	1,500–4,300	420–430
Pumped hydropower (utility scale)	5,400–14,000	900–1,400	6–10	80–82 (>13,000)	1,500–2,700	250–270
CAES (underground)	1,080–2,700	135.0	8–20	varies (>13,000)	1,000–1,250	60–125
Sodium-sulfur	300	50	6	75 (4,500)	3,100–3,300	520–550
Advanced lead-acid	200–250	20–50	4–5	85–90 (2,200–4,500)	1,700–4,900	425–950
Vanadium redox	250	50	5	65–75 (>10,000)	3,100–3,700	620–740
Zinc-bromide redox	250	50	5	60 (>10,000)	1,450–1,750	290–350
Fe/Cr redox	250	50	5	75 (>10,000)	1,800–1,900	360–380
ISO fist frequency regulation w/RE integration						
Flywheels	5	20	0.25	85–87 (>100,000)	1,950–2,200	7,800–8,800
Lithium-ion	0.25–25	1–100	0.25–1	87–92 (>100,000)	1,085–1,550	4,340–6,200
Advanced lead-acid	0.25–50	1–100	0.25–1	75–90 (>100,000)	2,000–4,600	2,770–3,800
Storage for utility T&G grid support						
CAES (above ground)	250	50	5	varies (>10,000)	1,950–2,150	390–430
Sodium-sulfur	7.2	1.0	7.2	75 (4,500)	3,200–4,000	455–555
Vanadium redox	4–40	1–10	4.0	65–75 (>10,000)	3,000–3,310	750–830
Lithium-bn	4–24	1–10	2–4	90–94 (4,500)	1,800–4,100	900–1,700

Scoping the Business Case for Microgrids 237

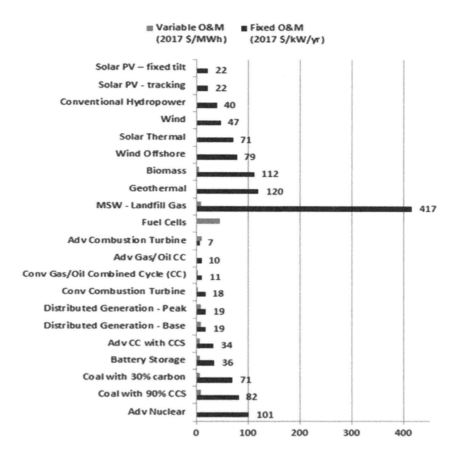

FIGURE 14.4 Operating and maintenance cost of electrical generating technologies (Source: Energy Information Administration [46]).

microgrid was being constructed at a cost of $2.55 million per MW (from Table 14.3), the total cost can be estimated at about $12.75 million. The value of electric demand (KW) and electricity generated (kWh) plus the operations and maintenance savings (avoided cost) would need to be estimated. Avoided costs might include the reduced costs of fuel for a diesel generator. Subsidies for the project might include renewable energy credits and possibly some type of loan subsidy. In addition to financing costs, other costs might include miscellaneous expenses and fees for services (e.g., load management or supplemental energy storage).

Assuming a 15-year financing term, a loan interest rate of 3.25%, and a down payment of $2.5 million, a preliminary cash flow analysis for the first ten years of the project might take the form shown in Table 14.5. The cash flow varies each year but is positive indicating that it is likely to be profitable for the developer. The cash flow begins to increase substantially after the first five years even though the loan subsidies are exhausted. Note that incremental utility rate increases and service costs are factored into the cash flow analysis. Over the term of the project, total cumulative

TABLE 14.5
Sample Microgrid Cash Flow Analysis.

Microgrid Project Year	15	1	2	3	4	5	6	7	8	9	10
Year	Testing	2021	2022	2023	2024	2025	2026	2027	2028	2029	2030
Annual KW Energy Savings	$200,000	$650,000	$682,500	$716,625	$752,456	$790,079	$829,583	$871,062	$914,615	$960,346	$1,008,363
Annual kWh Savings	$100,000	$400,000	$412,000	$424,360	$437,091	$450,204	$463,710	$477,621	$491,950	$506,708	$521,909
Annual O&M Savings	$0	$75,000	$77,250	$79,568	$81,955	$84,413	$86,946	$89,554	$92,241	$95,008	$97,858
Renewable Energy Credits	$0	$50,000	$50,000	$50,000	$50,000	$50,000	$50,000	$50,000	$50,000	$50,000	$50,000
Yearly Cash Inflows	**$300,000**	**$1,175,000**	**$1,221,750**	**$1,270,553**	**$1,321,502**	**$1,374,696**	**$1,430,238**	**$1,488,237**	**$1,548,805**	**$1,612,062**	**$1,678,131**
Annual Payment	$0	$868,805	$868,805	$868,805	$868,805	$868,805	$868,805	$868,805	$868,805	$868,805	$868,805
Loan Subsidy	$0	($100,000)	($100,000)	($50,000)	($50,000)	$0	$0	$0	$0	$0	$0
Miscellaneous Expenses	$0	$25,000	$25,000	$25,000	$25,000	$25,000	$25,000	$25,000	$25,000	$25,000	$25,000
Service Fees	$0	$20,000	$20,400	$20,808	$21,224	$21,649	$22,082	$22,523	$22,974	$23,433	$23,902
Yearly Cash Outflows	**$0**	**$813,805**	**$814,205**	**$864,613**	**$865,030**	**$915,454**	**$915,887**	**$916,329**	**$916,779**	**$917,239**	**$917,707**
Net Cash Flow for Year	$300,000	$361,195	$407,545	$405,939	$456,472	$459,242	$514,351	$571,908	$632,026	$694,823	$760,423
Cumulative Cash Inflows	$300,000	$1,475,000	$2,696,750	$3,967,303	$5,288,804	$6,663,500	$8,093,738	$9,581,975	$11,130,780	$12,742,842	$14,420,973
Cumulative Cash Outflows	$0	$813,805	$1,628,011	$2,492,624	$3,357,654	$4,273,108	$5,188,995	$6,105,323	$7,022,103	$7,939,341	$8,857,048
Cumulative Net Cash Inflows	$300,000	$661,195	$1,068,739	$1,474,678	$1,931,150	$2,390,392	$2,904,743	$3,476,652	$4,108,678	$4,803,501	$5,563,924

Total Project Cost	$12,750,000
Downpayment	$2,500,000
Financed Investment Cost	$10,250,000
Rate of Financing	3.25%
Financing Term (Years)	15
Annual Utility Rate Increase	5.00%
Annual O&M Savings Increase	3.00%
Annual Service Cost Increase	2.00%
Total Net Cash Flow (Life of Project)	$5,563,924

cash inflows exceed cash outflows including the initial down payment. This example microgrid financial analysis provides a strong cash flow and indicates that the project developers have a proposal that would appeal to investors. However, it takes just over five years for the initial down payment to be recaptured. Depending on the configuration of the microgrid and the type of equipment used, the life of the equipment might exceed the financial term of the project. At that point in time (in the example after 15 years) financing costs are eliminated and the net cash flow of the project increases. Note that this example analysis does not capture any savings for reduced downtime due to outages.

Once a lifecycle cost analysis is performed, the microgrid might have a negative projected cash flow. In such instances, it is difficult to obtain financing for the project. Increasing the down payment on the project is an option if the project developer is willing to assume greater risk and has additional funds available to invest. Negative cash flow can also be overcome by meeting additional customer needs, reducing operating costs, offering additional supplementary services, obtaining subsidies, or accessing incentives.

These are the sort of findings that are common for microgrids that use renewables. After the costs of development are amortized and the costs for financing are paid, there is no continuing cost for fuel and no risk associated with fuel pricing variability. Therefore, the costs for the electricity generated by the microgrid tend to decline considerably.

SUMMARY

Microgrids have the ability to eliminate waste, increase power efficiency, provide power to critical facilities in the event of a blackout, provide service and support to the bulk power grid, offer price response to lower wholesale power prices for customers, lower emissions, and serve as a catalyst for economic development [39]. Not all benefits of a microgrid system are financial. Facilities that require uninterruptable power systems typically rely on emergency backup generators that may not perform consistently due to infrequent use [39]. In such cases microgrids can provide substantial financial benefits if the risks can be quantified.

There is vast potential for financial incentives to target emissions and procure the consumer-driven changes necessary for most mitigation strategies. Other types of financial support that may assist municipal microgrid developers include energy bonds such as qualified energy conservation bonds (QECBs), taxable bonds, and tax-exempt bonds plus any applicable tax credits [6].

Identifying quantifiable benefits is an important part of the lifecycle analysis. The potential financial benefits include: 1) enhancing grid resilience and stability; 2) improving energy security; 3) matching power quality to end-user requirements; 4) providing ancillary services to the grid; 5) lowering carbon footprints by incorporating renewables; and 6) enabling market participation of DER and energy storage [36]. Whether such incentives are created by the market or supported by governments, they can be remarkably effective tools in the fight to forestall and avoid global climate change. In particular, this chapter illustrates the capacity of the private sector to promote carbon reduction strategies through investment and participation in fiscal endeavors, such as new technologies, CDM mechanisms, ESCO/ESPC programs, and nascent carbon markets. Energy savings performance contracts are a means of implementing and financing energy efficiency upgrades. ESPC projects are subsidized with utility savings and other cost avoidance. An energy service company guarantees savings that result from the facility upgrades. ESPC projects that improve the efficiency of the use of carbon-based fuels or use substitute forms of energy provide carbon mitigation potential.

Microgrids represent an integral part of GHG reduction strategies; they contribute to mitigation efforts by using renewable energy and managing energy flows. Cap-and-trade systems attach a monetary value to carbon emissions and create a quantity ceiling for plan emissions. Thus, the generators who reduce GHG emissions further than required can sell the remaining allowance. Likewise, carbon offsets—the promotion or maintenance of biological processes or natural features that sequester carbon—can also be translated into a monetary value.

Altogether, the potential for financial solutions to eliminate the threat of global climate change, stall the growth of GHG emissions, protect the environment, save money through energy-efficiency measures, and encourage the development of new industry is evident. Sequestration techniques are in the process of being developed and revised for widespread use in the U.S. CCS remains at the demonstration stage of the innovation process because of its high costs. The expense of CCS implies that countries with many large point source emitters (e.g., power plants and other industries) in close proximity to storage sites are the most likely candidates for cost-effective CCS programs. The financial restraints and possible risks involved in CCS procedures signify the need for government investment and regulation in developing technologies. This suggests the need for joint, concerted efforts among proponents in developed nations, where the appropriate technology, infrastructure, and geographic features make CCS a possibility. Evidence indicates that if carbon prices reached $75 to $185 per ton by 2020, then they would deliver substantial emission reductions by mid-century.

To assess the business case for microgrids, the scale and level of the proposed microgrid and its services must be identified. Microgrid development costs are known to vary by the location of the microgrid, the market segment, and other factors. In addition to the costs of generation and distribution assets, costs will likely include storage technologies. Goals of the business model include defining a project that offers a positive cash flow and can be financed. To establish a business case for a potential microgrid, the costs and benefits of the project must be estimated. This requires identifying the value proposition. For a lifecycle cost analysis, the operation and maintenance costs of the facilities must also be estimated. Projects with a sustainable positive cash flow have a greater likelihood of being financed.

Knowledge about the possible impact of fiscal decisions should encourage every stakeholder to act responsibly in their investment decisions and financial pursuits. People need to become more aware of how finances impact anthropogenic carbon emissions. This chapter offers a sense of empowerment and a dose of accountability in its underlying message that justifying a business case for microgrid development is an important part of the project development process.

REFERENCES

1. Dreessen, T. (2008, July 18). *Scaling Up Energy Efficiency Financing*. Paper presented at Wilton Park, UK.
2. Lazard (2017, November 2). Levelized cost of energy 2017-version 11.0. https://www.lazard.com/perspective/levelized-cost-of-energy-2017, accessed 24 November 2018.
3. GTM Research (2016, February). Integrating high levels of renewables into microgrids: opportunities, challenges and strategies. www.sustainablepowersystems.com/wp-content/uploads/2016/03/GTM-Whitepaper-Integrating-High-Levels-of-Renewables-into-Microgrids.pdf, accessed 16 January 2019.
4. National Renewable Energy Laboratory (2017). Annual technology baseline. http://atb.nre.gov.
5. U.S. Energy Information Administration (2018, March). Levelized cost and levelized avoided cost of new generation resources in the Annual Energy Outlook 2018, Table 1b. https://www.eia.gov/outlooks/archive/aeo18/pdf/electricity_generation.pdf, accessed 1 January 2020.
6. Microgrid Knowledge (2019). Community microgrids, a guide for mayors and city leaders seeking clean, reliable and locally controlled energy. https://microgridknowledge.com/.../guide-to-community-microgrids, accessed 13 January 2019.
7. Saury, F. and Tomlinson, C. (2016, February). Hybrid microgrids: the time is now. http://s7d2.scene7.com/is/content/Caterpillar/C10868274, accessed 22 January 2020.
8. Stern Review Report (2006). *The Economics of Climate Change*. Cambridge University Press.
9. Intergovernmental Panel on Climate Change (2007, November). Report on *Mitigation of Climate Change*, Chapter 8, Agriculture. Cambridge University Press: Cambridge, United Kingdom.
10. Massachusetts Institute of Technology. Clean water. http://web.mit.edu/12.000/www/m2012/finalwebsite/solution/econ.shtml, accessed 1 January 2020.
11. Stern Review (2007). The economics of climate change. http://www.rainforestcoalition.org/documents/Chapter25Reversingemissions.pdf, page 545, accessed 19 January 2009.
12. International Soil Tillage Research Organization (2007). Reversing emissions from land use change, chapter 25. *Stern Review*.
13. Dilling, L. and Moser, S. (2007). *Creating a Climate for Change*, chapter 26. Cambridge University Press: Cambridge, UK.
14. Wikipedia. Renewable energy certificate (U.S.). https://en.wikipedia.org/wiki/Renewable_Energy_Certificate_(United_States), accessed 24 November 2018.
15. Carbonify.com. What are carbon offsets and green tags? http://www.carbonify.com/articles/carbon-offsets-green-tags.htm, accessed 24 November 2018.
16. Global Environment Facility (2019). About us. https://www.thegef.org/about-us, accessed 17 January 2020.
17. Silveira, S. (2005). The Clean Development Mechanism (CDM), in Bioenergy – Realizing the Potential. https://www.sciencedirect.com/topics/engineering/clean-development-mechanism, accessed 24 May 2020.
18. Global Environment Facility (2007). About the GEF. https://www.thegef.org/sites/default/files/publications/GEF-Fact-Sheets-June09_3.pdf, accessed 24 May 2020.
19. Capoor, K. and Ambrosi, P. (2006). State and trends of the carbon market 2006. The World Bank, Washington, D.C.
20. Wikipedia (2020, May). Certified emissions reduction. https://en.wikipedia.org/wiki/Certified_Emission_Reduction, accessed 24 May 2020.

21. CDM Executive Board (2006, December 20). Clean Development Mechanism project design document form (CDM-PDD). https://cdm.unfccc.int/filestorage/Y/O/M/YOMB 0O8TNG9XC9KSOKT4NG8I5C55NZ/PDD.pdf?t=NkJ8cWF1YnBvfDD0rAJIT2E aHlK9tvcSqM7X, accessed 24 May 2020.
22. Shaw, F. and Shimamoto, S. (2018, November). A new frontier for solar – utility build-transfer agreements. *North American Clean Energy*, 12(6), page 10.
23. Morris, J. (2018, November). Separating hype from reality – blockchain-based peer-to-peer energy trading. *North American Clean Energy*, 12(6), page 8.
24. Roosa, S. (2019). Implementing sustainable development agendas using public-private partnerships. *International Journal of Energy and Environmental Planning*, (1)1, pages 66–75.
25. Fainstein, S. (1994). *The City Builders: Property, Politics and Planning in London and New York.* Oxford: LBlackwell.
26. Clarke, S. and Gaile, G. (1998). *The Work of Cities.* University of Minnesota Press: Minneapolis, Minnesota.
27. Schneider, M. (1989). *The Competitive City.* University of Pittsburgh Press: Pittsburgh, Pennsylvania, pages 12–26.
28. Blakely, E. (1994). *Planning Local Economic Development: Theory and Practice.* Sage Publications, Inc.: Thousand Oaks, California.
29. Woodruff, E. and Roosa, S. (2019). Financing and performance contracting, in chapter 25, *Energy Management Handbook* (Roosa, S., editor), pages 671–694. The Fairmont Press: Lilburn, Georgia.
30. Hansen, S. (2006). *Performance Contracting: Expanding Horizons.* The Fairmont Press, Inc.: Lilburn, Georgia.
31. United States Federal Energy Management Program. Fact sheet: energy savings performance contracting. http://www1.eere.energy.gov/femp/pdfs/espc_fact_sheet.pdf.
32. Public Works Magazine (2009, February). *Kilowatt Killers*, page 16.
33. California Energy Commission (2018, August). Microgrid analysis and case studies report. CEC-500-2018-022. https://ww2.energy.ca.gov/2018publications/CEC-500-2018-022/CEC-500-2018-022.pdf, accessed 23 January 2020.
34. Grimley, M. and Farrell, J. (2016, March). Mighty microgrids, ILSR's Energy Democracy Initiative. https://ilsr.org/wp-content/uploads/downloads/2016/03/Report-Mighty-Microgrids-PDF-3_3_16.pdf, accessed 24 January 2020.
35. Trabish, H. (2017, February 23). Pushing for scalability, utilities look to new hybrid models for microgrid deployment. https://www.utilitydive.com/news/pushing-for-scalability-utilities-look-to-new-hybrid-models-for-microgrid/435992, accessed 23 January 20.
36. Reilly, J., Hefner, A., Marchionini, B., and Joos, G. (2017, October 24). Microgrid controller standardization – approach, benefits and implementation. *Conference Presentation at Grid of the Future.* Cleveland, Ohio.
37. Miret, S. (2015, February 25). How to build a microgrid. http://blogs.berkeley.edu/2015/02/25/how-to-build-a-microgrid, accessed 23 January 2020.
38. New York State (2018). Microgrids 101. https://www.nyserda.ny.gov/All-Programs/Programs/NY-Prize/Resources-for-applicants/FAQs#why-invest, accessed 22 November 2018.
39. Mansouri, H., Jalali, M., and Sabouri, H. (2018, June 7). A novel optimal electricity pricing method in microgrids based on customers' participation levels. http://www.cired.net/publications/workshop2018/pdfs/Submission%200239%20-%20Paper%20(ID-20749).pdf, accessed 24 January 2020.
40. Mosher, J. (2018, November). Clean peak standard. *Distributed Energy*, pages 48–49.
41. Sandia National Laboratories (2014, March). The advanced microgrid: integration and interoperability. Sandia Report SAND2014-1535.

42. Wood, E. (2016, April 26). What does a microgrid cost? https://microgridknowledge.com/microgrid-cost, accessed 2 February 2020.
43. Giraldez, J., Flores-Espino, F., MacAlpine, S., and Asmus, P. (2018). Phase I microgrid cost study: data collection and analysis of microgrid costs in the United States. National Renewable Energy Laboratory, NREL/TP-5000-67821, page 6.
44. Electric Power Research Institute (2010, December). Electricity energy storage technology options, a white paper primer on applications, costs and benefits, EPRI, Palo Alto, CA. http://files.energystorageforum.com/Epri_White_Paper.pdf, pages xxiii-xxv, accessed 23 January 2020.
45. U.S. Department of Energy (2011, August 30). DOE microgrid workshop report. https://www.energy.gov/oe/downloads/microgrid-workshop-report-august-2011, accessed 19 January 2019.
46. U.S. Energy Information Administration (2018, February). Cost and performance characteristics of new generating technologies. *Annual Energy Outlook 2018.*

15 It's Back to the Future with Microgrids

Microgrids are not a new technological solution for generating and delivering electricity. Historical models for today's microgrids are found in several of the oldest electrical generation systems. The focus of microgrids is how electricity delivery systems can be scaled and configured to meet specific local requirements. When we think about the future of microgrids we consider the wide variety of ways there are to generate electricity. Microgrids in the past often used combinations of fossil fuels and renewable energy (RE). The same is true today.

IT'S ALL ABOUT PROVIDING ELECTRICITY

Electrical generation is driven by customer desires. What customers want is electricity that is inexpensive, reliable, resilient, and readily available. What they don't need is electricity that when generated creates unmanageable wastes and causes unreasonable environmental issues. Electrical generation has a history of creating unforeseen environmental problems. It is well known that fossil fuels (coal, oil, and natural gas) when used to generate electricity create pollution in our atmosphere, in bodies of water, and on the land. When combusted, they also cause greenhouse gas (GHG) emissions. Large central atomic power plants have unresolved issues with nuclear wastes which have little potential to be adapted for reuse. In fact, wastes are typically stored on-site and the costs for storage over thousands of years is unknown. Not all environmental pollution is manageable and it often has unforeseen impacts. These Rankin cycle generation systems also consume vast amounts of fresh water. Therefore, our need for electricity is tempered by the way electricity is produced and the associated externalities of the generation processes.

Fossil fuel-fired generation solutions are not only environmentally problematic but also inefficient. Too much energy is lost as waste heat and during transmission. Combined heat and power (CHP) systems benefit from improved efficiencies. However, to be economically feasible plants must be co-located with facilities that have need for the heat being generated. When this is possible, both the costs for transmission infrastructure and the energy losses are reduced.

The opportunities to generate electricity have expanded. There are renewable energy technologies including hydropower, solar power, wind power, biomass and biofuels, and geothermal energy which offer solutions for electrical generation. Smaller and scalable package RE systems are available that no longer require central stations. Battery storage systems allow us to store electricity for use at a later time. These systems can be used in combination to overcome problems with intermittencies. While all electrical generation systems have their own identifiable externalities, RE generation typically creates much less atmospheric and water pollution.

Some renewables including hydropower, biomass, and geothermal energy sources are capable of producing baseload electrical power. Others including solar and wind power are viewed as being locally intermittent resources.

BRIGHT FUTURE FOR MICROGRIDS

Microgrids are a back-to-the-future solution and they offer great potential to provide electricity locally. In a world that has over a billion people without electrical power, they offer hope for a brighter tomorrow, one with the benefits of electrification. Supplying this growing demand for power solely with fossil fuel generation will exacerbate many problems that we are trying to resolve. The hope is found in renewable energy and energy storage systems. In combination, they appear to be among the best ways to provide power without negative externalities, unanticipated problems, and weighty subsidies.

The operation of microgrids offers potential advantages for customers and utilities that include improving energy efficiency, minimizing total energy consumption, reducing environmental impacts, improving supply reliability, plus providing network operational benefits such as loss reduction, congestion relief, voltage control, security of supply, and more cost-efficient electricity infrastructure replacement [1]. Many believe that local owners of microgrids are likely to make more balanced choices, such as between investments in efficiency and supply technologies [1]. Microgrid operators have the ability to coordinate the electrical generation assets and present them to the utility grid at a scale that is consistent with grid operations, thereby avoiding new investments for integrating decentralized energy resources [1]. This is due to the wide variability of potential microgrid configurations and components which can be operated from a central control system (see Figure 15.1).

There is a bright future for microgrids. The recent drop in the costs for renewable generation and storage technologies and the forecasts for continued declines offer hope for expanded markets. As costs for these generation systems continue to decline and their efficiencies improve, the markets for their applications broaden. Today, the lifecycle costs for land-based wind power, hydropower, geothermal, and some types of solar are competitive with coal.

CREATING SMART MICROGRIDS

The energy market in the U.S. is becoming more diversified as environmental regulations increasingly impact the central grids and consumers produce more electricity autonomously [3]. Future microgrids will require incremental improvements and upgrades to meet the reliability demands of their customers. Many people are producing their own renewable power and forming their own community grids [3]. For example, the state of California is the first to require that by 2020, all new residences must be constructed with either installed solar panels or a shared solar power system serving a group of homes.

Smart grid, based on the Energy Independence and Security Act of 2007, refers to the evolution of the electric grid toward a "modernization of the electricity delivery system so it monitors, protects and automatically optimizes the operation of

It's Back to the Future with Microgrids

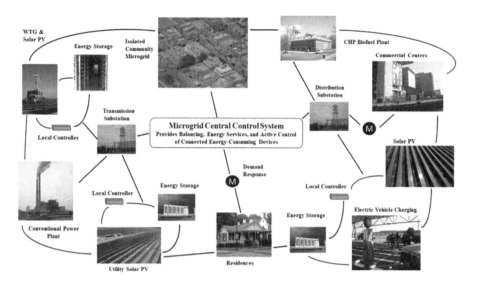

FIGURE 15.1 Topology of the microgrids of tomorrow. (Source: adapted from [2].)

its interconnected elements—from the central and distributed generator through the high-voltage network and distribution system, to industrial users and building automation systems, to energy storage installations and to end-use consumers and their thermostats, electric vehicles, appliances and other household devices" [4]. According Eger et al., "A smart microgrid generates, distributes and balances the flow of electricity to consumers, but does so locally. It aggregates and controls largely autonomously its own supply- and demand-side resources in low-voltage and even medium-voltage distribution grids" [5]. It is an electricity network based on digital technology that is used to supply electricity to consumers using two-way communications [6]. Smart grid devices transmit information that enables ordinary users, operators, and automated devices to quickly respond to changes in smart grid conditions [6].

Microgrids have been proposed as a novel distribution network architecture within the smart grid concept, capable of exploiting the benefits from integrating large numbers of small-scale distributed energy resources into low-voltage electricity distribution systems [1]. Smart grids will provide the consumer portals, technologies, data standards, and protocols that will enable the two-way communication of electronic messages between consumer-owned networks and intelligent equipment [4]. They provide multiple pathways to the development and implementation of diverse energy-related services (see Figure 15.2) [4]. Leveraging the ability to communicate with consumer-owned energy management systems and intelligent energy consuming equipment not only adds value but also allows for the decentralization of marketable services [4].

Microgrid applications operate using a consolidated control and energy management system with smart metering capabilities [7]. Advanced sensor technologies and data from metering systems analyze power flow data to identify opportunities and

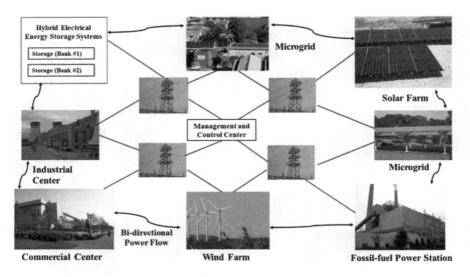

FIGURE 15.2 Diagram of smart grid electrical supply network with clusters of distributed energy resources. (Source: adapted from [8].)

optimize energy use while accounting for peak demand [7]. These smart technologies log and communicate asset performance in real time to improve decision-making processes and prevent costly unplanned maintenance events [7]. At the highest level, the smart grid configuration has the following components [1]:

- improved operation of the legacy high-voltage grid (e.g., by using synchro phasers)
- enhanced grid-customer interaction (e.g., by using smart metering and real-time pricing)
- new distributed entities that did not exist previously (e.g., microgrids and active distribution networks) [1]

The capabilities of the smart grid include more efficient operation, load management, demand response support, consistent reliability, premium quality power, the ability to self-repair systems, and consumer participation in grid operations [6]. The smart grid offers economic benefits by reducing costs and optimizing the benefits of energy services. As utility grids become more intelligent, microgrids can add value by providing services such as load leveling, reliability support, and demand response. The central grid structures of the future can be supported by clusters of distributed energy resources.

Development of Distributed Energy Resources

Driven by the need to supply electricity to remote regions, developing countries are taking the lead in DER development. Much of the development in remote locations is configured as microgrids. With increased electricity demand and lower technology costs, developing nations are now dominating the global transition to clean power. This reverses past trends when the world's wealthiest countries accounted for the bulk of renewable energy investment and deployment activity [9]. Consider the following trends identified in a recent report by Bloomberg NEF: 1) the majority of the world's new zero-carbon power capacity in 2017 (114 GW including hydropower and nuclear) was built in developing countries compared to 63 GW added in wealthier countries; 2) renewables accounted for the majority of newly added power-generating capacity (186 GW); 3) clean energy new-build deployment additions are growing fastest in developing nations, increasing 20.4% annually; 4) roughly 35 emerging markets have held reverse auctions for clean power-delivery contracts with the estimated levelized cost of electricity for wind and solar substantially below $50/MWh in many developing nations (e.g., in Mexico $21/MWh for solar PV, in India $41/MWh for wind power); and 5) as of the end of 2017, 54 developing countries had invested in at least one utility-scale wind farm and 76 had received financing for solar projects [9]. The majority of the funding for these projects was from local sources.

EMERGING ELECTRICAL GENERATION TECHNOLOGIES

There is a wide assortment of new electrical generation technologies that are becoming available. Many of these emerging technologies are just now being tested and deployed. All have a local or regional focus and offer various levels of economic feasibility. These technologies will be deployed to a greater extent as their costs decline and their feasibility becomes more proven. Many provide opportunities to generate electricity for microgrids. A sample of these technologies are discussed.

Wave Energy Systems

The uneven heating of the Earth's surface drives the world's winds creating surface stress in bodies of water and causing waves to form. Wave energy is a form of hydropower whose kinetic energy is from surface waves or pressure fluctuations below the surface. Wave energy is sometimes classified as a type of ocean energy, yet wave energy can also be harnessed using waves in large lakes (e.g., Lake Superior in North America or the Caspian Sea in Asia). Systems can use offshore, near shore, or shoreline devices. The electrical energy generated is then harnessed at the coastline or transferred to shore by undersea electrical cables. The world's first commercial wave farm using systems that ride on the waves opened in 2008 at the Aguçadora Wave Park in Portugal. Others are located off of the coasts of Oregon, Scotland, and Spain.

Wave energy can be harnessed by float or buoy systems that use the rise and fall of ocean swells to drive hydraulic pumps which stroke an electrical generator or dielectric elastomer generator to make electricity [10]. The most common categories of wave energy systems include terminators, attenuators, point absorbers, and overtopping devices [11]. There are also devices called capture chambers. These devices enlist shore-based oscillating water columns to use the in-and-out wave motion to enter a column [10]. They direct air through a constricted space to turn a turbine, filling the column with water as the wave rises and emptying it as the wave recedes [10]. In the process, air inside the column is compressed and warms, creating energy the way a piston does (see Figures 15.3a and 15.3b) [10]. Attenuator wave energy converters float on the ocean's surface or may be partially submerged [11]. As they "ride" on the waves, the wave passes the attenuator allowing energy to be harvested [11].

Another way to harness wave energy is to bend or focus the waves into a narrow, tapered channel, concentrating the waves and increasing their power and size. The waves can then be channeled into an elevated catch basin or used directly to spin turbines. Gravity causes water to exit the basin and electricity is generated using hydropower technologies.

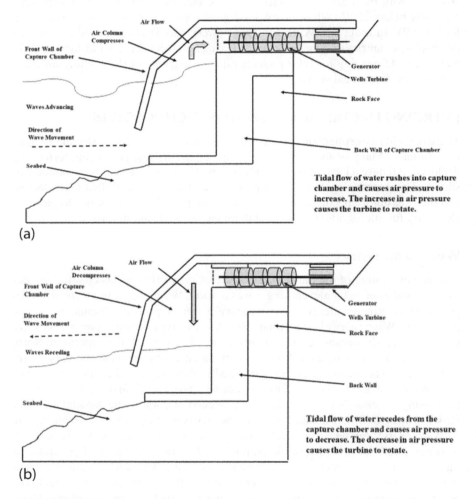

FIGURE 15.3 (a) Cross section of oscillating water column wave capture chamber system showing waves advancing (Source: graphic based on [12]); (b) cross section of oscillating water column wave capture chamber system showing waves receding (Source: graphic based on [12]).

Other ways to capture wave energy are being studied. Some of these devices under development are placed underwater, or anchored to the ocean floor, while others ride on top of the waves. Wing wave systems work with the concept that water particles orbiting under a shallow water wave follow elliptical paths [11]. As depth increases, the horizontal movement of the water particles' rotation remains the same while the vertical movement decreases. This causes the motion of the water near the ocean floor to oscillate horizontally. The wing waves capture this horizontal motion to move the wings of the system, ultimately pumping a working fluid to a generator via a hydraulic ram [11].

TIDAL POWER

Ocean mechanical energy is driven by tides which move due to the gravitational pull of the moon and the sun [13]. Despite being tiny in size compared to the sun, due to the moon's proximity to the Earth its gravitational force is 2.2 times stronger than that of the sun [14]. The tides follow the track of the moon during its orbit around the Earth, creating diurnal and ebb cycles on the ocean surfaces [15]. The predictability of tides, the vastness of tidal resources, and the development of complementary turbine and component technologies have led to increasing interest in exploiting tidal currents as an energy resource [15].

Electrical power can be generated using the tidal waters of bays, estuaries, or rivers. Unlike wind power resources which are unpredictable and solar energy which is unavailable when the sun is not shining, the movement of tidal water is both predictable and continuous. A tidal barrage is similar to a hydroelectric dam (see Figure 15.4). It is used to convert tidal energy into electricity by forcing the water

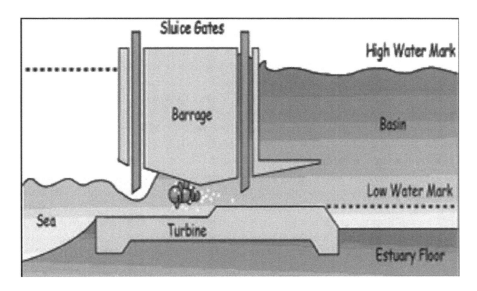

FIGURE 15.4 Cross section of a barrage retaining water in tidal basin [16].

through turbines, activating a generator [13]. They are among the oldest technologies used to generate hydropower.

A barrage system creates pressure head which is used to drive a turbine in a manner similar to a low-head hydroelectric dam. A sluice directs the water into a basin [16]. As the ocean level drops, gravity causes the water to flow back out into the ocean [16]. Barrage tidal power typically requires a difference between the high and low tides of 16 feet (4.9 m); there are about 40 regions in the world that have such conditions [15]. A benefit is that barrage systems can be designed to generate electricity when water is flowing both into and out of the basin or lagoon. The first major plant of this type in the world (240 MW) was constructed in 1966 at La Rance, France and produces 30 GWh annually [15]. Others can be found in the Bay of Fundy in Canada, Rance River in France, Sihwa Lake in Korea, and Kislaya Guba in Russia.

STIRLING-DISH ENGINE

A Stirling engine is a machine that can be used to provide either power or refrigeration. It operates on a closed cycle and uses a working fluid that is cyclically compressed and expanded at different temperatures. It is a closed-cycle regenerative heat engine with a permanently gaseous working fluid. Closed-cycle refers to a thermodynamic system in which the working fluid is permanently contained within the system, and regenerative describes the use of a specific type of internal heat exchanger and thermal store, known as the regenerator [17]. The use of the regenerator differentiates a Stirling engine from other closed-cycle hot air engines [17].

The Stirling engine when combined with a solar concentrator system more effectively applies existing proven technologies. This configuration uses solar concentrators to provide the heat that drives the Stirling engine [18]. Like other solar technologies, specific efficiencies are low and a substantial collection of solar energy is required. Despite having moving components, the key advantage of this system is its simplicity.

Dish-engine is the oldest of the solar technologies, dating back to the 1800s when solar-powered, Stirling-based systems were first demonstrated [18]. Modern technologies that use these engines began developing in the late 1970s but were not scalable. Since the 1990s, prototype demonstrations have attempted to fine-tune the technology, with the goal of developing 5 kW to 10 kW units for distributed power and 25 kW systems for utility-scale applications. In 2005, two major California utilities signed power purchase agreements totaling over 800 MW; these became the first commercially operational dish-engine power plants [18].

Dish-engine technology uses a parabolic dish reflector to concentrate direct radiation onto a receiver connected to a power conversion unit. Because of their small focal region, dish-engine systems track the sun on two axes—azimuth-elevation and polar [18]. The concentrators are typically configured to approximate an ideal shape by using multiple spherically shaped mirrors supported by a structural truss. Due to a short focal length and the need to focus solar rays directly at the receiver, dish-type systems require support structures that prevent vibration caused by variable wind conditions [18]. A receiver transfers solar energy to a high-pressure working gas, sometimes helium but usually hydrogen. Stirling engines convert heat to mechanical power in the same manner as conventional engines [18]. Power is produced by the expansion and contraction of the solar-heated working gas which drives a set of pistons and a crankshaft [18].

Dish-engines are attractive because of their high efficiency, modular design, and comparatively flexible siting requirements compared to other solar thermal technologies that require large expanses of flat land area [18]. When combined with solar technologies, constraints on wide-scale deployment include the cost of the truss assemblies that use steel and aluminum, the availability and cost of specialized dish-shaped mirrors, and the cost to maintain the engine [18]. However, other workable renewable energy applications might involve combining geothermal resources to provide the heat needed for the process. Geothermal or hot spring locations have naturally occurring temperature differentials. Placing the hot chamber of the Stirling engine inside a hot spring would be a means of gathering the heat necessary to keep the engine running, as long as a cooling system is maintained [19]. Geothermal heat does not require any specific location for utilization, since at certain depths the Earth maintains a constant temperature [19]. Geothermal Stirling engines could be used in areas without easy access to the grid, and function as either heat pumps or as generators [19].

A design-build project was launched to create a microgrid at Tooele Army Depot. The microgrid plan involved incorporating the existing generation sources at the depot: a Stirling engine-based solar array, two wind turbines, and a diesel generator [20]. The renewable resources were sized to have the capacity to power the entire 24,000 acre (9,712 hectare) installation; diesel generation was used for periods of low renewable power availability during grid outages [20]. To integrate the generation sources and control power distribution, a microprocessor-based system of relays, communications switches, and firewalls was used; it was among the first to be based fully on traditional SCADA systems [20].

Electricity Generated Using Hydrogen

Hydrogen can be used to generate electricity for microgrids. While it is the most abundant element in the universe, the key problem is that most hydrogen used today is derived from fossil fuels such as methane (CH_4), propane (C_3H_8), and butane (C_4H_{10}). As an alternative, hydrogen can be manufactured using renewable energy and stored. According to NREL, "…electrolysis is a process that uses electricity produced from renewables to split water into hydrogen and oxygen. The hydrogen can function as an energy storage medium, effectively storing RE until a fuel cell or engine converts it back to electricity" [21].

One promising methodology is to use any available excess electricity to produce hydrogen via electrolysis. This hydrogen could be used for vehicles as a mobility fuel (e.g., those using proton exchange membrane fuel cells) or to generate electricity, especially to improve capacities, provide load leveling, or meet peak demand requirements [21]. The first large hydrogen-fueled plant in the U.S. is planned to replace the 840 MW coal-fired Intermountain Power Plant near Delta, Utah. It will be a model for future similar grid-connected plants. The electricity generated by the plant is to be transmitted to Los Angeles; the project is driven by the need to reduce GHG emissions [22]. The plan for the plant is to install turbines capable of combusting a mixture of 30% hydrogen and 70% natural gas incrementally increasing the hydrogen ratio to 100% by 2045 [22].

Mitsubishi Hitachi Power Systems has announced a project in the Netherlands to convert a 440 MW natural gas combined-cycle gas turbine (CCGT) plant to 100% hydrogen combustion spearheading a new industry to make hydrogen fuel common across Europe and in the U.S. [23]. An important environmental benefit is that a CCGT plant using natural gas produces about 65% less CO_2 than an equivalently sized retiring coal-fired power plant [23]. When paired with a 50% gas and 50% renewable sources, the CO_2 reduction increases to 85% [23].

The vision for microgrids involves local generation of electricity using renewable energy resources. Excess power, perhaps generated by from wind and solar power, would be directed to an electrolysis facility and use fresh water to create hydrogen (H_2) and oxygen (O_2). The hydrogen would be piped to storage and used for electrical generation. Alternatively, H_2 could also be recombined with captured CO_2 (perhaps from a combined-cycle natural gas plant) to produce a synthetic natural gas for power plants or transportation applications [21]. Renewable hydrogen has a future in electrical generation as costs are projected to decline from $2.30–$6.80 per kg to $1.40–$2.90 per kg by 2030; however, only 253 MW of renewable H_2 projects have been deployed over the last two decades [22].

SMALL-SCALE NUCLEAR REACTORS

Uranium is a fossil fuel. It is often classified as a clean energy source when used for nuclear power processes that generate electricity. This is because GHGs are not emitted when electricity is generated. When we think of nuclear plants we focus on utility-scale generation facilities. There are other options that are often overlooked.

Small nuclear reactors have potential to power microgrids in the future. Advanced small modular reactors (SMRs) are currently being developed in the U.S. which can be used for power generation, process heat, desalination, or specific industrial applications [24]. They vary in output from a few to hundreds of megawatts and use light water coolants or non-light water coolants such as a gas, liquid metal, or molten salt [24]. They can be installed in multiples and sized to match the electrical load requirements of a university campus, small town, or military base.

SMRs offer advantages that include their small size (1/100 that of a traditional reactor), variable output capabilities, reduced capital investment, ability to be sited in locations not possible for larger nuclear plants, and the ability to incrementally add power [24]. Unlike large nuclear plants, SMRs can be constructed within the ten-mile (16-km) radius of their service areas. SMRs can operate either independently or be connected to an electric grid, allowing them to power a campus or small community in the event of grid failure [25]. SMRs can store over two years' of their fuel requirements on-site, allowing them to maintain power after extreme weather events or other threats to the electric grid [25].

The next generation of small reactors might use thorium as fuel and molten salt as a heat sink to provide electricity with no nuclear waste [26]. Integrated smart grid capabilities both in distribution and consumption can ensure that nuclear microgrids would manage the ebb and flow of demand and production far more efficiently than today's power generation infrastructure [26].

While fusion reactor demonstration projects are likely decades away, smaller nuclear reactors are more likely to be available sooner. Integral Molten Salt Reactor (IMSR) technology is a small-scale and modular technology that is being explored by a company located in Ottawa, Canada. IMSR uses molten salt as both the fuel and the coolant and operates on a variety of nuclear fuels including spent nuclear waste [27]. IMSRs can work in combination with RE facilities such as solar and wind plants to produce continuous utility-grade, fossil fuel-free energy with no carbon footprint [27]. IMSR plants can also operate under ambient pressures making them much safer than conventional nuclear plants. They are not subject to the potential of radioactive gas explosions and there is no risk of meltdown upon failure [27]. An example is the NuScale SMR. This technology uses a natural circulation light water reactor with the reactor core and helical coil steam generators located in a common reactor vessel in a cylindrical steel containment (see Figure 15.5) [28]. The reactor vessel containment module is submerged in water in the reactor pool, which is also the reactor's heat sink and located below grade [28]. The reactor building is designed

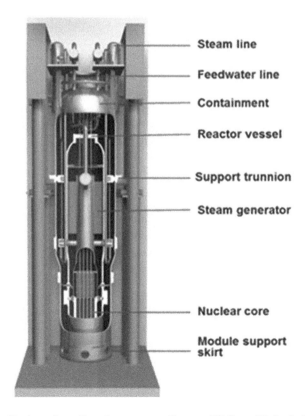

FIGURE 15.5 Design of small nuclear reactor. (Source: NRC.gov/NuScale [28].)

for 12 SMRs. Each SMR has a rated thermal output of 160 MW and electrical output of 50 MW each, yielding a total capacity of 600 MW for all twelve SMRs [28].

While there are regulatory issues, permitting, and construction delays that are associated with large nuclear reactors, these hurdles might be overcome with smaller versions of nuclear technology. Package nuclear energy systems though expensive are only a fraction of the cost of large-scale nuclear plants. There are unresolved environmental concerns with SMR deployments, yet they are more manageable. However, regulatory structures are not in place to enable wider application of package nuclear systems. In most U.S. states enabling legislation is not in place and several outlaw the use of nuclear power regardless of the technology used.

The availability and comparatively lower cost of natural gas and renewable generation such as wind power and hydropower make nuclear plants the less competitive option. Renewable generation technologies are proven, regulatory legislation is usually in place, and the time required for development is less. They remain the more cost-effective solutions to providing electricity with clean energy resources. Regardless, the first micro-nuclear fission fast-reactor (rated at 1.5 MW) in the U.S. might be operational between 2022 and 2025. If completed as proposed for Idaho Falls, Idaho, it will provide power equivalent to that required by about 1,000 homes, use recycled nuclear waste for fuel, and operate for about 20 years without being refueled.

PLASMA-ARC GASIFICATION

Plasma gasification is a waste-to-energy (WTE) process. It is an emerging technology which processes landfill waste to extract commodity recyclables and convert carbon-based materials into fuels [29]. It provides a means of producing renewable fuels while achieving near zero-waste byproducts. Plasma-arc processing has been used in the past to treat hazardous waste, such as incinerator ash and chemical weapons, and convert them into non-hazardous slag [29]. There are plasma gasification plants operating in Canada, Japan, India, and elsewhere.

Plasma gasification is an extreme thermal process using plasma that converts organic matter into a synthesis gas (syngas) which is primarily composed of hydrogen and carbon monoxide [30]. The process has been compared to passing municipal waste materials through a lightning bolt. A plasma torch powered by an electric arc is used to ionize gas and catalyze organic matter into syngas [30]. The heat from the plasma arc (over 8,000°C), and the intense ultraviolet light of the plasma, result in the complete cracking of tar substances and the breakdown of char materials, creating a synthetic gas [31].

This process uses most feedstocks, often municipal solid waste, in an initial gasification process, followed by clean-up, heat recovery to produce steam, and then generation of electricity. The clean synthetic gas exiting the plasma converter is then cooled and conditioned by scrubbers before being used directly by reciprocating gas engines or gas turbines to generate renewable energy [31]. Residual heat is recovered from the process to be used in CHP applications within the process [31]. Waste byproducts include vitreous slag and metals.

The electricity generated by plasma-arc systems is greater than that needed by the process so the excess power can be exported. Other benefits of the process include low environmental emissions and no generation of methane gas. The process does not discriminate in type of feedstock, and the ability to generate a synthetic gas. Since over 99% of the feedstock is used in the process, the economics of these plants are most favorable when there are high costs and tipping fees associated with municipal landfills.

DEVELOPING APPLICATIONS FOR MICROGRIDS

Recent technological applications will impact the development of microgrids. This section provides a sample of emerging non-generation technologies, artificial intelligence, and wireless energy transmission that will impact the microgrids of the future. While cost reduction for microgrid development has in the past focused on the hardware and technologies associated with generation, transmission, and storage devices, the future will focus to a greater extent on reducing the costs associated with microgrid management and operations.

ARTIFICIAL INTELLIGENCE

One of the key features of microgrids is the fact that most require custom design which creates the need for specialized engineering expertise. This design process can be selective and cumbersome. Another feature of microgrids is their tendency to grow and develop iteratively over a period of time. Microgrids tend to establish themselves on a small scale and grow as additional generation or storage assets are incrementally added to the system. While some generation technologies are adaptable others are not. Variable loads must be managed. There is a developing market for microgrid control systems that can identify irregularities, consider the various ways to respond to events, and take corrective actions in real time.

Artificial intelligence (AI) and machine learning that can assist in resolving these issues and applications for microgrids are on the horizon [32]. Introspective Systems and Israel-based Brightmerge are incorporating software into a custom microgrid software platform that determines microgrid feasibility, and creates optimal design specifications for operational controls [32]. The feasibility study engine performs an algorithmic process that improves the speed and results of evaluating microgrid projects while the design optimization engine considers all the assets and variables associated with a specific microgrid project to configure an optimal mix of assets [32]. A pilot project is underway for a solar-plus-storage microgrid on Maine's Isle au Haut that will apply a form of this software [32].

AI can take advantage of data proliferation to provide predictive analytics and real-time insights into asset operations [33]. Already some energy providers are using AI to enhance clean energy resources as the lowest-cost energy option, increase their reliability and output, speed deployment, and provide improved services to their customers [33]. Through improved data analysis, AI applications are reducing the perceived risks to investors associated with clean energy, helping investors obtain greater returns on their clean energy investments, and expediting their deployment [33].

The variability of weather conditions causes wind and solar resources to be labeled as intermittent energy sources. Virtual power plants whose control centers use AI might begin to solve this problem. StatkrafIt, a company headquartered in Norway, uses a cloud-based AI platform to connect more than 1,500 wind, solar, and hydropower plants across Europe with electricity generation and electrical storage facilities [34]. This large virtual power plant located in Germany has a capacity greater than 12,000 MW [34]. Weather data, power generation forecasts, electricity production data, and market prices for energy are continuously fed into the VPP's data center, enabling Statkraft to match demand with supply in real time [34].

WIRELESS ENERGY TRANSMISSION

Wireless power transmission (also called inductive power transfer, or WiTricity) technologies are not a new phenomenon. In 1890, this technology was demonstrated by Nikola Telsa. It involves transmission of electricity to adjacent devices through the air and works based on the principles of magnetic induction transmission using an oscillating magnetic field. There are three main systems used for wireless electricity transmission: solar cells, microwaves, and resonance [35]. In an electrical device, microwaves are used to transmit electromagnetic radiation from a source to a receiver [35].

Wireless energy transmission (WPT) is commonly used to charge batteries. Similar to cell phone battery charging, some use magnetic induction charging techniques, which require that a mobile device be in direct contact with a charging device. Others use resonance charging, which allows a mobile device to be placed near the power source without the need to be in direct contact for charging [36]. Magnetic induction charging typically uses a transmitter coil and a receiver coil. Alternating current in the transmitter coil generates a magnetic field, which induces a voltage in the receiver coil [36]. Resonance charging is based on the same transmitter/receiver coil technology as magnetic induction, but it transmits the power at a greater distance [36].

Laser technology can be used to transfer power in the form of light energy, and the power converts to electric energy at the end of the receiver [35]. It receives power using different sources such as the sun, an electricity generator, or a high-intensity-focused light [35]. The size and shape of the laser beam are formed by optics. The transmitted laser light is received by the PV cells and converts the light into electrical signals [35].

Microwave power transmission (MPT) involves combining transmitting and receiving technologies. Basically, the MPT antenna array emits the power to the receiving antenna array. The transmitting array would be co-located at or near the source of generation. In the transmission section, the microwave power source generates microwave power controlled by the electronic control circuits [35]. The waveguide circulator protects the microwave sourced from the reflecting power which connects through a co-ax waveguide adaptor [35]. According to the signal propagation direction, the reducing signals separate by the directional coupler [35].

WPT can be used to transmit power generated by satellites (e.g., a solar PV station) perhaps in low geosynchronous orbit to receiving stations on the Earth's surface. These receiving stations in the future might power microgrids in remote locations. Other targets could be mobile platforms used to supply power in locations where there has been a major weather event that causes a local power outage.

CONCLUSION

The future of microgrids will be framed by the adoption of improved and more widely applied renewable energy resources. For the U.S., this means improving the flexibility of the nation's electrical grids to enable the capabilities of diverse renewables to be maximized. To become more flexible, it must have the ability to accept new electrical generation technologies and ride smoothly through changes in demand and supply [21]. Microgrids are a back to the future solution. To be successful, they must be capable of incorporating the capabilities of microgrids to supply excess power to the grid. With a more flexible electric system, more than 80% of total U.S. electricity generation could be supplied by renewables by 2050 [21]. Smart microgrids are the key to providing electricity more economically. There are a number of emerging technologies which may be used to generate electricity for future microgrids. They include wave energy systems, tidal power, the Stirling-dish engine, hydrogen, small-scale nuclear reactors, and plasma-arc generation.

One question that lingers on the horizon is whether or not the regulation of microgrids can evolve at a pace that prevents customers from defecting en masse from the grid creating a financial disaster for primary electric utility companies and causing equity and macroeconomic problems [32]. The inclusion of electric vehicles, demand response, and net meeting are game-changers [37]. Tomorrow's microgrids will likely use artificial intelligence, wireless electricity transmission, and other technologies.

REFERENCES

1. Berkeley Lab (2018). About microgrids. https://building-microgrid.lbl.gov/about-microgrids, accessed 2 January 2020.
2. Moran, B. and Lorentzen, M. (2016, March 22). Assessing the role of energy efficiency in microgrids. Presentation. Georgia Institute of Technology, Climate and Energy Policy Lab. https://aceee.org/sites/default/files/pdf/conferences/mt/2016/Lorentzen_Moran_MT16_Session3A_3.22.16.pdf, accessed 22 August 2019.
3. Bezdomny (2016, January 27). Microgrids paving the way for distributed energy. https://www.resilience.org/stories/2016-01-27/microgrids-paving-the-way-for-distributed-energy, accessed 2 January 2020.
4. Gellings, C. (2015). *Smart Grid Implementation and Planning.* The Fairmont Press: Lilburn, Georgia, pages 1–20.
5. Eger, K., Goetz, J., Sauerwein, R., Frank, R., Boëda, D., Martin, I., Artych, R., Leukokilos, E., Nikolaou, N., and Besson, L. (2013, March 31). Microgrid functional architecture description. FI.ICT-2011-285135 FINSENY, D3.3 v1.0.
6. Techopedia (2017). Smart grid. https://www.techopedia.com/definition/692/smart-grid, accessed 15 December 2019.
7. Black and Veatch (2017, June 13). Top 3 benefits of military microgrids. https://3blmedia.com/News/Top-3-Benefits-Military-Microgrids, accessed 3 December 2018.

8. Think. A green future for electrical networks. https://www.um.edu.mt/think/a-green-future-for-electrical-networks, accessed 22 August 2019.
9. BloombergNEF (2018, 18 November). Climatescope. Emerging markets outlook 2018. Energy transition in the world's fastest growing economies. http://global-climatescope.org/assets/data/reports/climatescope-2018-report-en.pdf, accessed 6 February 2020.
10. Ocean Energy Council (2018). What is ocean energy? http://www.oceanenergycouncil.com/ocean-energy/wave-energy, accessed 2 January 2020.
11. Wells, B. (2012, July 20). Wave energy systems – final report. http://my.fit.edu/~swood/Wave%20Energy%20Systems%20Final%20Report%202012.pdf, accessed 26 February 2020.
12. Wave power. https://sites.google.com/site/khc33toby/types-of-clean-energy-generation-systems/ocean-energy, accessed 2 January 2020.
13. Energy Systems and Sustainable Living (2016, March 17). https://ecoandsustainable.com/2013/03/17/wave-and-tidal/, accessed 2 January 2020.
14. Mazumder, R. and Arima, M. (2005). Tidal rhythmites and their applications. *Earth-Science Review*, 69(1–2), pages 79–95.
15. Gorji-Bandpy, M., Azimi, M., and Jouya, M. (2013). Tidal energy and main resources in the Persian Gulf. *Distributed Generation and Alternative Energy Journal*, 28(2), pages 61–77.
16. U.S. Department of Energy, National Renewable Energy Laboratory (2014). Ocean power. https://www.energy.gov/sites/prod/files/2014/06/f16/ocean_power.pdf, pages 210–215, accessed 2 January 2020.
17. Wikipedia. Stirling Engine. https://en.wikipedia.org/wiki/Stirling_engine, accessed 19 November 2019.
18. Caunt, G., Baker, S., and Roosa, S., editors (2018). *Energy Management Handbook*. Roosa, S. (ed.), pages 489–490. The Fairmont Press: Lilburn, Georgia.
19. Chen, H., Czerniak, S., Cruz, E., Frankian, W., Jackson, G., Shiferaw, A., and Stewart, E. (2014, March 28). Design of a Stirling Engine for electricity production. https://web.wpi.edu/Pubs/E-project/Available/E-project-032814-103716/unrestricted/Stirling MQP_Final_2014.pdf, accessed 19 November 2019.
20. Burns and McDonnell (2019). Army depot microgrid. https://www.burnsmcd.com/projects/army-depot-microgrid, accessed 19 November 2019.
21. National Renewable Energy Laboratory (2013, April). NREL leads energy systems integration. *Continuum Magazine*, page 4.
22. Roth, S. (2019, December 10). Los Angeles wants to build a hydrogen-fueled power plant. It's never been done before. *Los Angeles Times*. https://www.latimes.com/environment/story/2019-12-10/los-angeles-hydrogen-fueled-intermountain-power-plant?fbclid=IwAR3GX-Emn5b8VQkxYwUItCi52wfO-F8hAVgbw5F6LBS9ApuJaa7urxxannA, accessed 14 December 2019.
23. Browning, P. (2019, April 9). The energy industry's new focus on LNG, hydrogen and renewable technology is already enabling a change in power. *Forbes*. https://www.forbes.com/sites/mitsubishiheavyindustries/2019/04/09/the-energy-industrys-new-focus-on-lng-hydrogen-and-renewable-technology-is-already-enabling-a-change-in-power/#2073459f4198, accessed 16 December 2019.
24. Office of Nuclear Energy. Advanced small modular reactors (SMRs). https://www.energy.gov/ne/nuclear-reactor-technologies/small-modular-nuclear-reactors, accessed 23 August 2019.
25. Office of Nuclear Energy (2018, January 25). DoE report explores U.S. advanced small modular reactors to boost grid resiliency. https://www.energy.gov/ne/articles/department-energy-report-explores-us-advanced-small-modular-reactors-boost-grid, accessed 23 August 2019.

26. Rosen, L. (2013, April 7). The future of microgrids. http://hplusmagazine.com/2014/04/07/the-future-of-microgrids, accessed 23 August 2019.
27. Rosen, L. (2013, April 24). Energy update: Canadians in the nuclear business are borrowing and building new technology. https://www.21stcentech.com/energy-update-canadian-company-offers-nuclear-energy-alternative, accessed 23 August 2019.
28. U.S. Nuclear Regulatory Commission (2019, January 9). Design Certification Application – NuScale. https://www.nrc.gov/reactors/new-reactors/design-cert/nuscale.html, accessed 14 December 2019.
29. Dodge, E. (2017). Plasma gasification: clean renewable fuel through vaporization of waste. https://waste-management-world.com/a/plasma-gasification-clean-renewable-fuel-through-vaporization-of-waste, accessed 2 January 2020.
30. Wikipedia. Plasma gasification. https://en.wikipedia.org/wiki/Plasma_gasification, accessed 24 May 2020.
31. Advanced Plasma Power (2017). Process overview. http://advancedplasmapower.com/solutions/process-overview, accessed 16 September 2017.
32. Burger, A. (2019, September 5). Microgrids made easier – and smarter – with software that uses artificial intelligence. https://microgridknowledge.com/microgrids-transactive-energy, accessed 25 November 2019.
33. Francetic, A. (2019, August 29). Artificial intelligence pushes 'commoditized' wind and solar power into the money. https://www.greentechmedia.com/articles/read/artificial-intelligence-pushes-commoditized-wind-and-solar-power-into-the-m#gs.101idkh, accessed 25 November 2019.
34. Ziady, H. (2019, November 7). Renewable energy's biggest problem. *CNN Business*. https://www.cnn.com/2019/11/07/business/statkraft-virtual-power-plant/index.html, accessed 28 November 2019.
35. College, A. (2020, January 21). Wireless power transmission technology with applications. https://medium.com/@Aryacollegejaipur/wireless-power-transmission-technology-with-applications-219357ef8aa2, accessed 6 February 2020.
36. Mearian, L. (2012, September 11). Cutting the cable: wireless charging becomes a reality. https://m.computerworld.com/article/2492302/cutting-the-cable--wireless-charging-becomes-a-reality.html, accessed 6 February 2020.
37. Sanchez, L. (2019, June 24). A new era for microgrids (quoting Lilienthal, P.). *Distributed Energy*. https://www.distributedenergy.com/microgrids/article/21086075/a-new-era-for-microgrids, accessed 11 November 2019.

Index

A

Active balancing, 79
Africa, access to electricity, 157–158
Air pollution, 85, 89
    sources and types, 31
Alternative energy, 56, 219
Amersfoort, Holland, 43
Amman, Jordan, 208
Anchor load model, 164
Andasol solar power station, 120
Anti-islanding, 71, 87
Antiqua, 98
Applications layer, 204
Artificial intelligence, 257–258
Asheville, North Carolina, 49
Asset coordination, 83
Atlantic City projects, 228
    convention center, 228
    wastewater plant, 130–131
Atmospheric temperatures, 31

B

Bankymoon, 194
Barrage, 251
Battery management system, 79–81
Biomass energy, 54, 60–61, 95
    bio-fuels, 48–49
    biopower, 49
Bitcoin, 197
Black start, 192
Blockchain, 189–191, 193–194, 196–198, 225
    challenges and barriers, 197–198
    NXT, 194
BlockCharge, 196–197
Bonaire, 7
    Borrego Springs, 151, 154, 226
Bosnia and Herzegovina, 177–178
BREEAM, 44
Buildings, 42–47
    energy-efficient, 46
    net positive energy, 45
    residences, 46
    sustainable, 42–43
    zero net energy, 44–45
Build-transfer agreements, 224–225

C

Cap-and-trade programs, 221–223
    examples, 222–223
    how they work, 221
Capture chamber, 249–251
Carbon, 18–20
    capture, 218–219
    dioxide emissions, 20–23, 136
    financial instruments, 221
    footprint, 22, 205
    global emissions, 22–23
    low carbon economy, 28–29
    management, 24–25
    offsets, 223–224
    taxation, 217
    United States, 23
Certification Scheme, 176
Chicago Climate Exchange, 221
Cincinnati Zoo, 49
Clean Coalition, 153
Clean Development Mechanism, 223–224
Clean energy, 56
Clean Power Plan impact, 58
Climate Change Adaption Roadmap, 145
Climate disasters, 9–10
Coal plants, 27–28, 57–58, 67, 92
    clean coal, 65
    economics of, 27–28
Combined cycle plants, 59, 219–220
Combined heat and power, 4, 39–40, 60
    cogeneration, 39, 86
    renewable, 135–136
Communications layer, 204
Controller technologies, 232
Cyber-security, 62–64, 198

D

Developing countries, 63, 192–193
    incentives for, 193
Diesel generation, 59–60
Distributed energy resources, 3–4, 91, 98, 104, 109, 125, 189, 234
Distributed energy storage systems, 2–4
Distribution system operator, 173
Dynamic stochastic optimal control, 206

263

E

Edison, Thomas, 37, 39
    Appleton Edison Light Company, 37
    Illuminating Company, 37
Electric power, 53–65
    capacity defined, 94
    distribution, 64–65, 73–75, 203–204
        diesel generation, 59–60
    emerging technologies, 249–256
    generation system efficiencies, 78
    global access, 14
    grids, 46–48, 53
        defined, 53
        national, 54
        problems with, 62–64
        regional, 46–48
        regulation of, 65–66
    interconnections, 54
    levelized costs for generation, 220
    loads, 79
    outages, 64–65
    pricing, 232–233
    regulation, 65–66
    reliability, 63, 66
    sector comparisons, 29
    system structure, 55
Electric vehicles, 190
    charging, 49–50, 196–197
    energy backed currencies, 196
Energy Sector Management Assistance programs, 159
Energy service companies, 229–230
Energy storage systems, 79–81, 110–123, 129
    applications, 108–109
    battery, 116–119, 178–179
        comparative capabilities, 122
        discharge durations, 122
        lead-acid, 117
        lithium-ion (Li), 117–118
        management systems, 80
        sodium-sulfur (NaS), 118–119
        vanadium redox (VRB), 119
        zinc-air, 119–120
    conversion losses, 123
    compressed air storage, 114–116
    costs of different types, 236
    electrical, 116–117
    flywheel, 116
    hydrogen, 120
    pumped hydro, 111–114
    mechanical storage, 111–116
    microgrid storage systems, 121, 209–210
    roundtrip efficiency, 109, 117, 119
    thermal, 120–121
    traditional uses for, 109
    ultracapacitor, 79, 203
Energy supply, 190
    emerging generation technologies, 249–256
    security, 192
Energy trading, 194–195
Environmental issues, 65
    contaminates, 31–32
Ethereum platform, 193–194, 197–198
European Norms, 174
Externalities, 232–233

F

Financing microgrid projects, 219–229
    barriers, 197–198
    cash flow analysis, 237–239
    energy storage system costs, 236
    ESPC investment model, 229–230
    investments models, 159–160, 238
    market segment costs, 235
Fort Bliss, 6
Fort Bragg, North Carolina, 210
Fossil fuels defined, 54
Frequency control, 178–179
Fuel cells, 129, 134–135

G

Generation systems, 91–102
    comparative costs, 103–104
    complementary, 128
    gensets, 96
    overnight capital costs, 104
Geothermal energy, 100–102
    binary technology, 101
    flash technology, 101
    project example, 131–133
Ghana
    capacity and reliability, 157–158, 166
    impact of microgrids, 167–168
    investment models, 160
    microgrid policy and regulation, 165
    recommendations, 168–169
    Renewable Energy Act, 158
    solar PV microgrids, 164
    tariffs and rates, 167
    viability, 158–159
    Global Environmental Facility, 223
Global warming, 21–24
Gordon Bubolz Nature Preserve, 135
Gorona del Viento El Hierro, 113, 115
Grand Coulee hydro plant, 57
Grid-connected, 2
Grid Singularity, 194
Grids, regional, 46–48
Green buildings, 44
Greenhouse gases, 20–21
    contracts for emission avoidance, 218

Index

costs of reducing, 25–26
Green Power North America, 132
potential of reducing, 26–27
reduction strategies, 29–30
Regional GHG Initiative, 232
regulations and policies, 66, 85
Guaranteed service levels, 176
Guri power plant, 57

H

Hoover Dam, 58
Hybrid generation, 126–128
    advantages of, 128–129
    configuration, 137
    definition of, 126
    diesel and renewable, 129
    examples, 137
    financing, 221, 225
    fuel cells and renewable, 134–135, 148
    natural gas and renewable, 129–130
    nuclear power and renewable, 133–134
    solar and geothermal, 131–133
    solar and wind power, 130–131
Hydrocarbon Age, 18–22
Hydrogen power, 253–254
    vehicles, 50
Hydropower, 37
    barrage system, 251
    small systems, 94–95
    large systems, 56–57
    tidal power, 251–252

I

Iceland, 47
IEEE Standards, 44, 86–88
India, 25, 160, 163
Information sharing, 198
Information technology systems, 173
Inrush current, 177
Intelligent electronic devices, 205, 213
International Council on Large Electric Systems, 174
International Electrotechnical Commission, 174
International Energy Agency, 105, 174
International Energy Conservation Code (IECC), 43
Inverters, 83, 88, 166, 179–180, 185
Island mode, 2, 4, 11, 174–175

J

Johnson City, Tennessee, 96
Jordan solar PV plant, 208
Jouliette, 196

K

Kentucky Coal Museum, 28
Kodiak Island, 150
Kyoto Protocol, 21–23, 224

L

Landfill gas, 95–97, 135
Las Vegas, Nevada, 50
Ledgers, 191
LEED, 44
Les Anglais, Haiti, 209
Levelized energy cost, 220
Liquid fuel storage, 60
loads, 41, 53, 71–72, 75–81, 126
    asymmetric, 177
    categories of, 79, 81
    changes in, 126
    electric, 40, 126
    sharing, 128
    Long Island, New York, 152, 154
Louisville, Kentucky, 38, 50

M

Machine learning, 257
Manhattan Pearl Street Station, 37
Mannheim, Germany, 211
Marine vessels, 41–42
Measurement and verification, 217, 230
Metering, 202–203, 206
    advanced, 206
    smart, 204, 207, 209
Microgeneration Certification Scheme, 176
Microgrids, 34
    AC and DC, 41–42, 75–76, 126, 130
        advanced, 8, 202–203
        definition, 8, 202
        remote, 202
    architecture, 71–81, 84
        enterprise systems, 82
    benefits of deploying, 11–12, 201–202
    business case, 217–219
    campus, 148
    cash flow analysis, 235–239
    central control center, 80, 83
    challenges to development, 230–231
    community, 141–142, 150–153
        defined, 141
        drivers for, 141–142, 228–229
    components, 76–81
    configurations, 5, 7, 9, 81–82
        community, 144–145
        natural gas and renewables, 12–130
        operational, 82
        with renewable energy, 93, 137

connecting to the grid, 53–54
construction aspects, 176–177
container, 179–182
costs, 158, 235
definition, 2, 174
development costs, 231
disadvantages of, 12–13
distribution company, 151
examples, 147, 149, 151, 208–210
features of, 202–203
financial benefits, 239
future of, 246
generation costs, 233–235
government, 165–167
history of, 37–41
hydropower, 37
impacts, 164
    independent, 7
industrial, 146–147
interconnected, 7, 84
island, 174–175
    operational aspects, 175–176
    types, 175
isolated, 7–8, 158, 197
levels of, 234
load types, 79, 81
local service, 74, 144–145, 174–175
operational characteristics, 76, 79, 178–179
maintenance costs, 12–13, 233–234
managing, 211–212
marine, 41–42
military, 145–147
mobile, 8, 144, 179–182
perspectives on, 161–162
power management systems, 78–79
power sources, 76–77
private, 165–167
regulations and standards, 84–88
renewable, 93–94, 105–106
    hybrid renewable generation, 128
smart, 175, 204–206, 213, 246–248
    benefits of, 204–205
    examples, 207–211
    technologies, 206–207
    topology, 247
solar in Ghana, 164
standards, 86–88, 176, 179
    sustainable, 210–211
    types, 6–8, 143
utility distribution, 147–148, 151
virtual, 8, 148–150
Microwave power transmission, 258
Milligrids, 6
Mill towns, 40
Mini-grids, 142
Moapa River Indian reservation, 99

Mobile power plant, 179
Model City Mannheim, 211–212
Mojave, California, 100
M-Pesa, 193
Municipal solid waste, 102–103

N

Nanogrids, 42–46, 174
Natural gas plants, 58–59, 95
Netherlands, 43, 48
Network development, 182–184
Nuclear, 60–61, 254–256
    design of SMRs, 255
    fuel assembly, 61
    fusion reactors, 255
    internal molten salt reactor, 255
    nuclear power and renewable, 133–134
    plants, 61
    power, 60–62
    small modular reactors, 254–256
    submarines, 41–42

O

Ocean thermal energy conversion, 94–95
Off-grid network, 176–177
Offset aggregator, 221
Ohm's law, 39
Operation and maintenance, 233–234
    advantages of microgrids, 246
Overland Park, Kansas, 211

P

Paris Agreement, 23–24
Peer-to-peer trading, 190, 194–195, 225
    performance contracts, 229–230
Picogrids, 174–175
Pinellas County Plant, 102–103
Planning solutions, 48
    electrical distribution, 53–55
    military bases, 145–147
    principles, 177–178
Plasma-arc gasification, 256
Point of common coupling, 3, 77
Policies, 212–213
    development concerns, 206
    for large microgrids, 231
    power, 209–211
        bulk supplied, 53
        distribution and transmission, 53, 55
        layer, 204
            management systems, 71–72, 78–79
        outages, 9, 64, 204
        traditional sources, 56
Public-private partnerships, 226–229

Index

Pumped storage installations, 114
PURPA, 85–86

Q

Quality issues, 11–12, 174–176, 185

R

Regional Greenhouse Gas Initiative, 222
Reliability, 166
Renewable energy, 54
    biomass, 95
    certificates, 223
    credits, 223
    defined, 54
    EU Directive, 192
    municipal goals, 142–143
    power generation, 91–101
    solar, 97–100
    zones, 193
Resiliency, 6, 8, 11, 14

S

Saint Thomas Island, 47
Sandia National Laboratories, 81, 212
Santa Rita Jail, 135, 149
Security issues, 113, 146, 190
Sensor systems, 205–206
Smart contract, 191
Smart Energy Power Alliance, 202
Smart grids, 203, 246–248
    deployed projects, 209–210
    diagram of, 175, 248
    integrating renewables, 210–211
    technologies, 206–207
Smart meters, 248
Smart plug, 196
Software decision support, 182–183
    HOMER, 209
    schematic for EMS integration, 214
    worksheets, 182–183
SolarCoin, 195
Solar energy, 97–100
    high-temperature collectors, 97–98
    photovoltaics, 98–100, 211
        utility scale, 100, 208
    rooftop collectors, 43, 46–47
    thermal, 97–98
Solid waste treatment, 102–103
Southern Exhibition, 38–39
Stabilization triangle, 30
Stand-alone power, 2–3
Stillwater plant, 131–132
Stirling-dish engine, 252–253
Subjective evaluation method, 164

T

Tariffs, 167, 190
Tehachapi Pass, 101
Three Gorges Dam, 57
Tidal power, 251–252
Tradable renewable certificates, 223
Transactional models, 189–190
TransActive Grid, 194
Transportation systems, 48–49
    hydrogen vehicles, 50
    railways, 38, 48

U

Uninterruptable power supply, 239
United Nations World Food Program, 193
United States
    carbon dioxide emissions, 20–24
    Clean Air Act, 30–32, 85
    Energy Regulatory Commission, 66
    municipal renewable goals, 142–143
    NAOA, 10
    primary energy consumption, 92
    PURPA, 85–86
    standards, 31–32
    statutory regulations, 212–213
University of Bridgeport, 134
University of California San Diego, 135
University of the Sunshine Coast, 121

V

Virtual power plants, 148–150
Vodafone, 193
Voltage fluctuations, 175

W

Waste-to-energy, 102–103
Wave energy systems, 249–251
    capture chamber system, 250
Weather events, 9–11
Wind power, 100–101
Wireless energy transmission, 258–259

Y

Yangtze River, 57

Z

Zero-carbon, 249
Zero-emissions, 56
Zero net energy, 44–45, 210
Zhejiang Juhua chemical plant, 224